The
Science Class
You Wish You Had ...

The Science Class You Wish You Had...

The Seven Greatest Scientific Discoveries in History and the People Who Made Them

**David Eliot Brody
and Arnold R. Brody, Ph.D.**

A Perigee Book

Figures and diagrams prepared by James H. Brown and David E. Brody. Figures 13-2 and 16-7 are from electron microscope photographs taken by Arnold R. Brody.

Reprint of *The World Through the Eyes of a Child* courtesy of John Kuzma. Copyright ©1996 John Kuzma; used by permission.

A Perigee Book
Published by The Berkley Publishing Group
200 Madison Avenue
New York, NY 10016

Copyright © 1997 by David Eliot Brody and Arnold R. Brody, Ph.D.
Book design by Irving Perkins Associates
Cover design by Elizabeth A. Sheehan/Charles Björklund
Cover illustrations by Elizabeth A. Sheehan/J. Brown

First edition: August 1997

Published simultaneously in Canada.

The Putnam Berkley World Wide Web site address is http://www.berkley.com

Library of Congress Cataloging-in-Publication Data

Brody, David Eliot.
    The science class you wish you had— : the seven greatest scientific discoveries in history and the people who made them / David Eliot Brody and Arnold R. Brody.
        p. cm.
    "A Perigee book."
    Includes bibliographical references and index.
    ISBN 0-399-52313-8
    1. Discoveries in science—History.   I. Brody, Arnold R.
II. Title.
Q180.55.D57B76   1997
500—DC21   96-45097
                                                                              CIP

Printed in the United States of America

10 9 8 7 6 5 4 3 2

*Dedicated to our families
and to the memory of our father,
the pioneer*

Contents

Part Five: Evolution and the Principle of Natural Selection

Part Six: The Cell and Genetics

Part Seven: The Structure of the DNA Molecule

Epilogue: The Synthesis

Foreword

In this book's view, science can be likened to a beautiful tapestry woven with different threads that have common features, but also demonstrate the extraordinary diversity of human thought in the creative process. It is an important book to read, whether the reader is completely ignorant of science or is a professional scientist. It combines the rare talents of two individuals of disparate backgrounds: a scientist (Arnold Brody) and a practicing lawyer (David Brody). Their breadth of knowledge, keen sense of social responsibility, and their gift for writing beautiful prose combine to reveal not only the course of scientific discovery, but the aspirations and the frailties of the participants in the context of the social and moral climate in which the important discoveries were made.

Seven pillars of scientific thought and accomplishment are discussed in this book—discoveries that encompass five hundred years—from the Renaissance to the mid-20th century. The authors describe these discoveries in a manner that makes science and the history of science clear and understandable. We learn that our planet is like a speck of dust in a vast desert; our Sun, around which we revolve, is a minor star among billions of others. Life may not have begun on the planet Earth; the vital substances in meteorites spinning through the vastness of space may have deposited the seeds of life during a fruitful period of the planet's formation from clouds of hot gas. We are indeed an evolutionary product of chance, a late event in the development of multicellular organisms. The complexity of form and function is not unique to humans but has evolved to allow us to reason. Yet reason and the rigors of scientific thought came late in human development, only within a fraction of the time it took for the appearance of the first hominid.

This book was written shortly before the advent of the second millennium. The last major discovery discussed here is the structure of DNA followed closely by the decoding principles showing that this structure is the repository of all the information encoding living material. This momentous event led to the development of molecular biology and enabled scientists to uncover a wealth of hitherto buried information.

During the past 50 years, mainly as a result of the largesse of the U.S. government, there has been an extraordinary increase in biological research, bolstered in a major way by the participation of physicists and chemists. One of the legacies of the Holocaust and the disruptions caused by World War II was the movement to the United States of dozens of Europe's most important scientists in biology and physics. The National Institutes of Health in Bethesda, Maryland, supported their efforts, encouraging post-doctoral training of M.D.'s and Ph.D.'s in an atmosphere conducive to creative thought and action. A remarkable group of scientists arose— well-trained with a broad perspective on biology and medicine. They filled American universities and medical schools. Western Europe and Japan, devastated by World War II, benefited by having their young scientists trained at American medical schools and the National Institutes of Health. The result has been an enormous cross-fertilization of ideas around the globe, stimulating what has become the most productive period in the history of physics and biology. Scientific research is now truly a global effort in which intellectual barriers are nonexistent and cultural differences have been minimized. Indeed, the global society of scientists—of every cultural and ethnic persuasion—epitomizes the true spirit of the quest for knowledge.

Success has brought, paradoxically, dramatic changes in the scientific establishment and the process of scientific discovery. Scientists in all developed countries are witnessing a leveling off or even a downturn in the amounts of public funds devoted to science. As this book illustrates, discovery is no longer in the hands of a few. Group science, spurred in part because of decreased funding, is the rule. As a consequence, discovery has become incremental rather than dramatically episodic. However, it must be emphasized that the problems faced in science are enormous. In my field alone—biological communication—there have been dramatic changes in perception and knowledge, which are the result of discoveries made by relatively few scientists. As a result of these basic discoveries, the complex nature of the communicating system in cells and organs has been revealed, and has brought into focus the daunting task of integrating such knowledge. Certainly this task will require a concerted, multi-discipline effort if we hope to understand how communication systems operate at the cellular and organ levels. Individual scientists, no matter how brilliant, cannot possibly provide the answers. Group effort may solve, perhaps, the most mysterious aspects of life—how the brain functions, the basis of the "mind," and how the process of communication within cells and organs dovetails with genetic information to secure a more sound basis for Darwin's Natural Selection theory.

Discovery in such fields will be incremental as scientists face the limits of knowledge that can be acquired from investigating complex systems. This does not mean that the "end of science" is approaching, as some have suggested. Truth may be ephemeral, but Science, the Flower of Mankind, will continue as long as the human spirit exists on this planet. Brilliantly, this book takes the reader through the sowing of this flower's seed, the story of its growth, the lives of the extraordinary individuals who nurtured it, and the magnificent blossoming that is science in the 20th century.

Martin Rodbell, Ph.D.
Scientist emeritus, Nobel Laureate
(1994) in Medicine or Physiology

Acknowledgments

This project would not have been possible without the assistance of a talented and generous group of people. First, our cousin, Stephen Karon, understood astronomy, physics, and electromagnetic fields and forces before he was 10 years old. He is now the director of the Museum of Science and Technology in Syracuse, New York. For many years, he has consulted and lectured extensively on a variety of science subjects, and was a part-time educator, working to enhance science literacy among the public. We thank him for his careful review of the entire manuscript and his valuable contributions. Kenneth and Bette Brody, both chemistry teachers and experts in the history of science, took time to read several chapters critically and provided numerous suggestions to improve them. They also directed us to useful sources.

In the early stages of our research, Joseph Broz, Ph.D., a physicist with broad expertise and with much insight into the relationship between science and society, shared with us his knowledge of physics, relativity, and the history of science, and directed us to a number of excellent sources. Robert C. Amme, Ph.D., professor of physics at the University of Denver for the past thirty years, spent many hours reviewing the drafts of Part Three (Relativity) and explaining that subject. David's colleagues Becke McGee and Brad McKim also read Part Three and gave us several constructive suggestions that were incorporated into the manuscript. In combination, these four people helped us present relativity in a manner that will hopefully bring the seemingly untouchable subject of space and time down to Earth for the reader.

We thank Peter Jordan for his comments on Part Five (Evolution), for providing his expertise in geology and plate tectonics, and for pointing out the notable scientific discoveries and contributions made by his Scottish countrymen.

David's wife, Susan, read the entire manuscript and provided many creative comments. We thank her for those comments as well as for her patience, support, and encouragement. Dr. Daniel Butterman made a valuable contribution to the discussion on DNA in Chapter 19 concerning methods to identify bacteria responsible for periodontal disease. Gary Sattin reminded us of the importance of relating the great discoveries to issues that affect people in society today.

We thank the Penrose Library at the University of Denver and its staff members for the use of the library's rich depository of science research materials. In addition, the University of Colorado (Norlin Library), the branches of the city of Denver library system, and the Denver Museum of Natural History were very cooperative in providing their extensive and comprehensive resources and information.

We owe much thanks to our agent, Elizabeth Kaplan of the Ellen Levine Literary Agency in New York, who made this entire book possible by her early and accurate advice on the approach and direction it should take, and by finding the right publisher to bring it to fruition. Our editor at Perigee Books, Suzanne Bober, was highly supportive of this project from the beginning of our association with her. We greatly appreciate her comprehensive and thoughtful editing of the manuscript, and the expert manner in which she guided it through all the stages in the publishing process. The copy editor, Kim Mohan, performed masterful work in pointing out numerous words, phrases, and concepts in need of clarification or change. We thank him not only for his excellent copyediting skill, but also for his insight into the scientific and historical details.

Finally, Bonnie Rothschild did an excellent job of typing and revising the many drafts of the manuscript that were necessary over the years during which it was in the making. We thank her for the quality of her work and her promptness and reliability in always meeting our deadlines, often resulting in her working all hours of the day and all days of the week.

David E. Brody
Arnold R. Brody

Introduction

Can the average person really understand science? Does the average person want to know about science? Does science matter to us? The answer to these questions is a resounding YES!

For many of us, however, the mere memory of physics, chemistry, and biology classes in high school and college makes our eyes glaze over. We left the classroom with the belief that science was dull and abstract and virtually impossible for the average person to understand. Back then, it wasn't cool to understand science, and it seemed to have little immediate relevance to our lives. Yet as we matured and headed out into the world, we found ourselves face to face with sophisticated computers at work and frequent headlines about matters of science—mapping the human genetic makeup, cloning, test-tube babies, and the August 1996 discovery of the possibility of past life on Mars, to name a few. Suddenly, scientific knowledge has not only become acceptable, it has become a useful, essential, and inescapable part of our lives.

For some of us, our fascination with science began some time in the 1950's or 1960's, when the Soviet Union launched Sputnik or when Neil Armstrong set foot on the moon—striking evidence of mankind's ability to apply scientific knowledge to accomplish an extraordinary goal. For others, all it took to become interested in science was getting out of high school or merely witnessing the unending series of new scientific achievements and inventions that occurred during the 1970's, 80's, and 90's: Venus landings, fiber optics, reading the DNA code, black holes, space stations, microchips and computers, microsurgery, the Space Shuttle, heart transplants, artificial heart, superconductivity, the discovery of other solar systems, and much more.

You don't have to be a theoretical physicist to be awed by space exploration or curious about whether there's life on Mars or how the universe began. You don't have to be a biochemist to have an interest in the fundamental processes of life. It's impossible *not* to be curious about such matters. Scientific knowledge and discoveries are much too interesting and profound to be left only to scientists.

The Seven Greatest Scientific Discoveries in History . . .

All those phenomenal technological achievements and inventions we've witnessed and experienced over recent decades have a common thread: *They became possible as a result of the basic scientific discoveries in physics, chemistry, and biology (and their subdisciplines) made over the last four centuries.* This book identifies the greatest of those discoveries—seven findings so profound that almost everything else that humankind understands in science is based upon them:

1. Gravity and the Basic Laws of Physics
2. The Structure of the Atom
3. The Principle of Relativity
4. The Big Bang and the Formation of the Universe
5. Evolution and the Principle of Natural Selection
6. The Cell and Genetics
7. The Structure of the DNA Molecule

These are the discoveries of *what is*, not what has been invented or what has been modified or developed by mankind, such as the Space Shuttle, insecticides, or the discovery of penicillin. This is *not* an examination of microchips or computers. It *is* a look at how the universe and all life forms came into existence, how it all works, and what it's made of, from the smallest elementary particle to the most enormous and distant galaxy.

We do not ask or answer the question of *why* the universe or life exists. As the famous theoretical physicist Stephen Hawking stated, that would require knowing "the mind of God." And we do not ask what *should* be, since that's also outside science's domain. We can leave such questions to the philosophers.

The selection of these particular discoveries was based on a combination of criteria, including opinions of scholars and experts in science and the history of science. These seven discoveries form the foundation that supports the enormous body of scientific knowledge that's been built. Without an understanding of physics, there would be no Venus landings or Space Shuttle. If we did not know the structure and function of the atom, nuclear power plants and the threat of nuclear war would not exist. If the principles of genetics had not been discovered, farm production and the world's food supply would be greatly diminished. Without a thorough understanding of the DNA molecule, cures for Parkinson's disease, sickle-cell anemia, and hemophilia would not be on the horizon. Each breakthrough marked a new epoch in scientific knowledge, and often brought with it monumental ethical and philosophical debate and controversy. These discoveries not only unlocked the intellectual and material riches of the modern world, they've had an enormous impact on our daily lives.

The first four of these seven discoveries deal primarily with physics in its broadest sense and with astronomy. The last three deal with biology, including its related fields and subdisciplines (such as medicine and biochemistry). The matter and the concepts involved in these discoveries are impossible to see with the naked eye (as with the atom, the cell, and DNA), or are unthinkably abstract (in the case of gravity and relativity), or they embody and reflect an enormous event (the Big Bang), or a process of unimaginable duration (evolution). Because these discoveries can't readily be seen, or are so abstract, or cover an enormous space or time span, they conflict with our everyday experiences—our own common sense. As a result, neither the scientific community nor the general public realized the full significance of these discoveries at the time they were made; each discovery acquired its greatness only after withstanding intense scrutiny.

In each instance, to reach a full understanding of how the physical universe and life on this Earth were created and developed or evolved to their present state, we had to give up some perceptions and assumptions about science and life—what we thought was common sense. Today, the greatest obstacle to such full understanding is the so-called "wisdom" of the past. We're plunked down "here and now" in the 20th century with all the myths and half-truths that we've dragged with us from prior centuries, amid the complex social and cultural context that has developed over the last few thousand years. That body of beliefs and culture originated well before *any* great scientific discoveries—before anyone had the knowledge to answer profound questions about the creation of the universe and life, before we discovered the Big Bang or the principle of evolution, and before we recognized the clear distinction between concepts based on creativity or our fertile imaginations and concepts based on reality.

One has to approach the study of science with an open mind. You cannot fully understand the concept of gravity until you realize it is more than a matter of semantics to distinguish between an object *falling* and being *pulled* to the ground. The same force that keeps our feet on the ground causes the Earth to revolve around the sun. Accepting the underlying principles of science will change your fundamental views of life, the universe, and the human race. For example, understanding evolution presents an entirely new perspective on the battles and conflicts throughout history that were the result of physical and cultural differences between people—a perspective that tends to knock down barriers.

. . . And the People Who Made Them

This book also describes the triumphs and tragedies, lives and motives of the extraordinary people involved in those great discoveries. We discuss and recognize the work of hundreds of people, but give the primary credit to ten of them.

1. *Gravity and Physics*
   ISAAC NEWTON

2. *Atom*
   ERNEST RUTHERFORD
   NIELS BOHR

3. *Relativity*
   ALBERT EINSTEIN

4. *Big Bang*
   EDWIN HUBBLE

Their attitudes, personalities, and sheer determination, in combination with their ideas, led to these discoveries. For example, in 1543 Nicolaus Copernicus reluctantly published his book stating that the Sun, not the Earth, is the center of the universe. This idea touched off a revolution in thought that carried with great force and controversy into the next century. Galileo, the Italian mathematician, astronomer, and physicist, spent years attempting to prove the truth of Copernicus's theory. When Galileo put that proof in print in 1632 in *Dialogue on the Two Chief World Systems*, the Church determined he'd violated its prohibition against teaching Copernican views, and announced the book to be heresy. Ultimately, Galileo was forced to retract his words publicly and vow never to teach Copernican theory again or risk being tortured and burned alive at the stake.

Isaac Newton, the father of physics, led the world into the next scientific era. He was born in 1642, the year Galileo died. Newton's father had also died earlier that year, and the infant was left by his mother in his grandmother's care. Years later, with the university closed during the Great Plague of London, Newton retreated to his secret world at Woolsthorpe, England, where he invented calculus and conceived the idea of universal gravitation. Yet he didn't publish it for over twenty years. Newton's work was finally revealed in 1687 in his book the *Principia*, considered to be the greatest single scientific intellectual feat in history. As with any fundamental change in thought, this new idea called gravity was only accepted slowly. "You have not explained why gravity acts," they accused. So, like Copernicus, Galileo, and others before and after, Newton had to fight another battle in the war for rational thought.

In the mid-1800's a monumental battle was emerging in the field of biology, where Charles Darwin became the commanding general. His proposition was unique and dealt with an issue that had never before been the subject of the scientific method of investigation: the concept of life itself, its origin and evolution—biology and organic matter, not the remote lifeless orbs studied by Copernicus, Galileo, and Newton. Because of the emotional elements of Darwin's subject, this English gentleman found himself the focal point of the greatest controversy of the 19th century, one that continues unabated into the 21st century, from the Scopes monkey trial in 1925 to today's heated confrontations involving creationism versus evolutionism.

In many cases the social and political forces that these scientists faced influenced and often inhibited recognition of the importance of their discoveries. Science didn't make these people famous. Rather, these are the people who made science. This is history as well as science, for the two are inextricably linked, bound together with the economic, political, military, and religious strands that tie civilization into a coherent picture.

Yes, science can be fascinating. These seven discoveries have now, in our lifetime, culminated in the most incredible and pervasive scientific and technological revolution that could be imagined. Whether we approve of it or not, we're swept up in that revolution and the resulting culture—unless you live in a cave. Even then, you could have HBO and a cell phone. So, yes, not only is science fascinating, it matters to us because it is our lives. "The present epoch is a major crossroads for our civilization and perhaps for our species," wrote Carl Sagan in *Cosmos*. "Whatever road we take, our fate is indissolubly bound up with science. It is essential as a matter of simple survival for us to understand science." Knowing science means knowing life. It means feeling more comfortable with our everyday lives, and using science and technology to accomplish goals. Science is part of our culture and heritage. It is for the masses, not merely ivory tower intellectuals.

We are extraordinarily fortunate to be living in this exciting transitional period, unique in all of history—one in which we're going from ignorance to knowledge, from questions to answers, from wonder to understanding. As we approach the end of the century, a fog has lifted to show the people of the 1990's and the 21st century what those before us never knew. This is the objective of *The Science Class You Wish You Had:* By focusing on the greatest scientific discoveries in history and the people who made them, we hope to make science interesting and within every reader's grasp, to render science intelligible for the nonexpert, and to challenge personal myths, perceptions, and assumptions.

Once again, the answer to the first question posed earlier: *The average person really can understand the great scientific discoveries.* In this entire book there are only a couple of equations—such as Einstein's $E=mc^2$—and you won't be asked to use them. This book is about concepts and ideas, not algebra and calculus. Einstein himself said, "The whole of science is nothing more than a refinement of everyday thinking." Scientific discoveries are built on a foundation or framework supplied by mathematics, but words are all that are needed to understand them. By understanding the seven discoveries explored in the following pages and the role played by the scientists who made them, you will have accomplished a feat that would have astounded both you and your high school physics, chemistry, and biology teachers. This is the science class they should have been teaching. Finally, after many years, this book is what you needed. This is the science class you wish you had.

Part One

Gravity and the Basic Laws of Physics

In Focus

Ancient philosophies died slowly and stubbornly during the
Renaissance. As Copernicus, Tycho, and Kepler gradually proved that
Aristotle's and Ptolemy's theories about the universe were wrong,
people realized the Earth was not the center of the universe and that it
moves about the Sun. But the forces of reason came up against the
forces of politics and power—challenging the Roman Catholic
Church's strict Aristotelian teachings cost Giordano Bruno his life and
Galileo his freedom.

Then, in the latter part of the 17th century, in what is regarded as
the greatest single intellectual accomplishment in scientific history,
Isaac Newton discovered and described the law of universal gravitation
and the basic laws of classical physics. Every particle of matter
gravitationally attracts every other particle of matter with a force that is
quantified in Newton's mathematical formula. As massive bodies that
are made up of such particles, the Sun attracts every planet in our

solar system, and the Earth attracts the Moon. If you were to let go of this book, the Earth would *pull* on it with the same invisible force.

Universal gravitation, the laws of motion, and the other quantitative rules developed by Newton marked the beginning of modern physics and formed the paradigm on which much of modern science has been built. Reason triumphed, and the world was changed forever.

Chapter One

The Revolutions

"Any revolution . . . begins with some definite act, often meant to purify corrupt practices and restore what some conservative radical imagines as a pristine state of things."

CHARLES C. GILLISPIE
The Edge of Objectivity (1960)

"If you are a long-distance traveler . . . you can get your bearings by landmarks—mountains and rivers, buildings . . . But . . . the emptiness and homogeneity of the sea . . . drove sailors to seek their bearings in the heavens, in the sun and moon and stars . . . They sought skymarks to serve for seamarks. It is no wonder that astronomy became the handmaiden of the sailor, that the Age of Columbus ushered in the Age of Copernicus."

DANIEL J. BOORSTIN
The Discoverers (1983)

The Black Death Kills Millions and Threatens a Return to the Dark Ages

Perfect Circles and Crystal Spheres

In the summer of 1347, the most devastating plague ever suffered by mankind struck Europe, the bubonic and pneumonic plague caused by bacteria in fleas of rats. It became known as the Black Death, so-called because the victims' putrefying flesh supposedly blackened in the final hours before death. The plague originated in China and was transmitted to Europeans through a grotesque early version of germ warfare when a Kipchak army (a tribe of the Mongol Empire) attacked a trading post in Genoa, Italy, and used its giant catapults to lob plague-infested corpses over the walls of the town. In his book *The Black Death*, historian Philip Ziegler described the next deadly steps:

> As fast as the rotting bodies arrived in their midst, the Genoese carried them through the town and dropped them in the sea. But few places are so vulnerable to disease as a besieged city . . . Such inhabitants . . . realized that, even if they survived the plague, they would be far too few to resist a fresh Tartar onslaught. They took to their galleys and fled from the Black Sea towards the Mediterranean. With them traveled the plague.

Over the next four years, the disease spread in its first wave throughout Mediterranean ports in Sicily, North Africa, Italy, Spain, France, England, and finally all of Europe. The first symptoms were shivering, vomiting, headache, giddiness, intolerance to light, pain in the back and limbs, and eventually delirium. Quoting records from the period, Ziegler described the physical appearance and characteristics of the plague as boils "in the groin or armpits, some of which grew as large as a common apple" from where they "soon began to propagate and spread itself in all directions indifferently . . . a token of approaching death." When the lungs became infected, the victim died within two days. Other cases dragged out for months. Wrote Ziegler,

> Everything about it was disgusting, so that the sick became objects more of detestation than of pity; ". . . all the matter which exuded from their bodies let off an unbearable stench; sweat, excrement, spittle, breath, so foetid as to be overpowering; urine turbid, thick, black or red."

Efforts to check the disease included bonfires that were believed to disinfect the air, demonstrations of penitence, and persecutions of religious minorities. Twenty-five million people suffered those symptoms and perished—one-fourth of the population of Europe at the time, with some areas losing as much as 60 percent of their population. And this wasn't the end. There were five recurrences in the 14th century, and it would take two centuries before the European population would climb back to 100 million.

Nine hundred years earlier, after the fall of the Roman Empire in the 5th century, the Church of Rome had become the center of Western Christendom. By the year 1300 the Church had centralized its authority throughout Europe, becoming involved in secular elections, international conflicts, and all aspects of civic life. In Italy, the Church had successfully taken up the task of restoring order in city-states that had fallen apart politically and economically. This success enhanced papal authority, and by the time of the Black Death, the Roman Catholic Church had consolidated its power. Turning to religion for comfort and salvation, laymen became more pious, and morality more rigid. There were ugly political and physical attacks on European Muslims and Jews, as they were given the choice of Christianity or death. Most chose to convert.

The devastation of the Black Death brought with it the end of the prosperity and optimism that had begun to emerge in Europe in the preceding decades. Suddenly, there was a climate of fear and anxiety and a continuous series of social and political upheavals throughout the continent. Because much of the workforce had fallen victim to the plague, the fields became neglected. Serfs escaped from their masters' control and labor costs increased, leading to government efforts to control such costs, which culminated in the Peasants Revolt of 1381. Goods became scarce as the economy, largely dependent on agriculture for survival, fell into a medieval version of a depression. Economic and political chaos resulted, as Europe threatened to sink back into the Dark Ages that had enveloped Western civilization during the years 450–750.

At the time of the Black Death, the physical universe was defined by the principles laid down by the Greek philosopher Aristotle (384–322 B.C.) and the Egyptian astronomer Claudius Ptolemy (A.D. 100–170). Over the centuries, those principles had been tempered by the medieval interpretation adopted by the Church: God had set the heavens in perfect and eternal circular movement. Our world, made of the four elements (earth, air, fire, water), was at the center. Eight concentric crystal spheres made of an immutable substance, and the other heavenly bodies which were carried or supported by the spheres, comprised the heavens. One sphere carried the Sun, one carried the Moon, five separate spheres each carried one of the planets (other than Earth) known at the time (Mars, Mercury, Jupiter, Venus, and Saturn), and the eighth sphere carried all the stars. Material on Earth decayed and died, while the rest of the universe was perfect and unchanging.

Aristotle's and Ptolemy's ideas about the universe ultimately had an impact that extended well beyond a mere physical description of the natural world. An elaborate and entrenched set of cultural beliefs developed, based directly on this vision of reality. Astrology was central to this body of thought. For centuries astrologers— and the kings, queens and emperors who relied on the astrologers' advice—looked upward to determine the location, "influence," and "character" of these heavenly bodies for everyday guidance. In turn, the lives of millions of people ruled by the leaders who reigned throughout much of the medieval period were affected by Aristotle's and Ptolemy's view of the universe.

"Planet"	English	French	Italian	Spanish
Sun	Sunday	dimanche	domenica	domingo
Moon	Monday	lundi	lunedi	lunes
Mars	Tuesday	mardi	martedi	martes
Mercury	Wednesday	mercredi	mercoledi	miércoles
Jupiter	Thursday	jeudi	giovedi	jueves
Venus	Friday	vendredi	venerdi	viernes
Saturn	Saturday	samedi	sabato	sábado

Even the arbitrary compartmentalization of time into seven-day groupings is the direct result of this ancient concept of the heavenly bodies. Beginning during the Roman Empire of two thousand years ago, each day was dedicated to one of the seven "planets," though these included the Sun and the Moon, and excluded the

Earth. The specific order of the days of the week was based on the "influence" of each planet on worldly affairs, not their distance from the Earth. Thus institutionalized by astrology, our days of the week are named after the "planets" dating back to ancient Rome. As shown above, this connection remains obvious in English, and even more so in other languages.

This artificial cluster of seven days, not tied to any regular movement of the heavenly bodies, was a result of imagination. Though it was an attempt to find regularity, and included the idea that humans are influenced by invisible forces that act over great distances, the entire approach to figuring out the universe lacked any basis in fact.

Superstition and astrology ruled. Magic, witchcraft, and alchemy were popular. There was no science.

Commercial Needs Spark the Scientific Revolution

Johann Müller and the Epicycles

By the early 15th century, Europe began showing signs of an impending transformation. Indeed, the legacy of the plague was an incentive to climb out of that dismal era as quickly as possible and bury it in the past. It set the stage for a bright new chapter—the Renaissance, the rebirth of Europe, characterized by the rediscovery of classical literature, the rebirth of art, and great interest in all intellectual endeavors. The Renaissance included the Scientific Revolution, the age of worldwide geographical exploration and unprecedented efforts to develop technology aimed at facilitating economic and commercial enterprises.

Italy was first to fully recover from the Black Death. Since their country was ideally located for commercial trade between Europe and the Near East, Italians had been heavily engaged in world trade since the early 1300's, importing spices, perfume, silk, and other commodities from the Orient and brokering these items to northern European countries. As a result, Italy had developed sophisticated managerial systems, banking practices, and financial expertise that paved the way for the nation's renewed commercial preeminence in the 1400's. Florence was regarded as the world leader in international banking and trade. Italy's remarkable economic recovery following the medieval period also led to that nation's preeminence in scientific inquiry. Mathematics was used in construction, navigation, cartography, and surveying. Beginning in the 1400's, geometry, trigonometry, and algebra were refined and applied to a greater extent than ever previously imagined. Sixteen centuries earlier, a few Greek philosophers had suggested the Earth was round—a view that had been buried deep in history all through the Middle Ages. But thanks to Paolo Toscanelli (1397–1482), the Italian cosmographer who provided Columbus with the map that guided him on his first voyage, and the new generation of explorers, which included Columbus, Vespucci, and Magellan, man began to realize the Earth was not a disc primarily comprised of Europe and Asia. It was a sphere largely covered with water, and all of the distant lands might be habitable.

With the use of mathematics in everyday life and the recognition that the Earth is round, ideas about the universe, including the movement of the Moon, the Earth, and other planets, began to change. If a starting point for that change could be identified, it would be 1463, not in Italy, but in Germany, where an astronomer named Johann Müller (1436–1476) wrote a book called the *Epitome* (published posthumously in 1496) in which he noted weaknesses in Ptolemy's geocentric, or Earth-centered, theories expressed in *The Almagest*. Müller was particularly critical of Ptolemy's convoluted "epicycles," or small loops within each planet's orbit, that were needed to account for the movements of those planets in a universe in which the Earth stood still at the center. Although *The Almagest* had been written over 1300 years earlier, it remained the authority on how the universe functioned—one of the great symbols of the unwavering reverence for the ideas espoused by Aristotle and Ptolemy. Müller's work was the first step toward renouncing the belief that the Earth was unmoving, immovable, and the center of the universe, and set the stage for a revolution in astronomy.

Copernicus Topples Geocentricity and Dethrones Mankind

But Where Is the Wind?

We now know that it takes the Earth 365 days, 5 hours, 48 minutes, and 46 seconds to make one trip around the Sun. The ancient Egyptians calculated a year as 365 and *one-fourth days*—but that's 11 minutes, 14 seconds too long. Thus, when Julius Caesar adopted the Egyptian calendar in the 1st century A.D., it led to a gradual disparity between the dates on the calendar and the seasons, so that by the mid-15th century the Julian calendar was ten days off. Farmers and navigators noticed the problem. Also, the Church had based Easter on the vernal equinox, which had originally been fixed at March 21, but was now occurring on March 11 because of the accumulating inaccuracy.

In 1475, Pope Sixtus IV had asked Johann Müller to undertake a study to determine the exact cause of the disparity. But he failed in that effort, for he lacked the data, and it was virtually impossible to accumulate such data under the prevailing astronomical system of Aristotle and Ptolemy with the Moon and Sun circling the immovable Earth, and the eight crystalline spheres eternally rolling around the Earth. The continued adherence to Aristotle and Ptolemy was the bedrock of social stability, yet it was also an impediment to an accurate Easter and the very birth of science.

In the meantime, Nicolaus Copernicus (1473–1543) was growing up in Torun, Poland, the youngest of four children of a prosperous merchant. Copernicus had a broad range of interests and talents and studied at several schools in Italy—astronomy at Bologna, medicine at Padua, and law at the University of Ferrara. He later lectured in various universities throughout Europe on mathematics, astronomy, and medicine, but spent most of his life as canon of Frauenburg, Poland, with his education and career under the auspices of the Catholic Church, the same as most scholars of the period.

In 1506 he began to develop an astronomical system based on his own observations and calculations on the movement of heavenly bodies. Like Müller, Copernicus soon recognized that Ptolemy's theories of a geocentric universe were not consistent with these observations. On the other hand, Copernicus knew that Ptolemy's writings were consistent with the Bible and that it does not appear the Earth is rotating on its axis or revolving around the Sun. The concept of gravity was not known yet, and Copernicus had no convincing argument to counter the universally accepted Aristotelian explanation that the reason objects fall to Earth is because matter is "naturally attracted" inward to the center of the universe, that is, the Earth. Copernicus didn't attempt to explain why objects did not "fall" off the Earth and toward the Sun if the Sun is in fact the center of the universe. Moreover, few were prepared to consider a theory that displaced Earth and humankind from the hub of the universe. The unassailable truth was that God put us in the center and on the largest object, as vanity, fear, and the Bible demanded this correlation of location and size to importance.

When recruited by the Pope's secretary in 1514 to investigate the same perplexing problem of the inaccurate calendar that had eluded Müller, Copernicus was faced with the dilemma of trying to explain his observations without refuting the popular and sacred notions of the day. Dissension between Catholics and Protestants made the Church more vigilant about maintaining its teachings, thus increasing the chance that Copernicus would be charged with impugning the Church's authority. As a result, he declined the Church's invitation and told them he could not explain the reason for the discrepancy until the relationship among the Earth, the Sun, and the Moon was more fully understood. He continued to secretly develop his theory while holding various governmental and ecclesiastic posts—from 1519 to 1521 he aided the reconstruction of Ermland (now the Warmia region) in northern Poland, served as commissary for the Ermland diocese and provided medical help to the community, and in 1522 he presented a plan to reform the local currency.

Convinced of the accuracy of his observations and encouraged by friends, Copernicus finally went public in 1530 when he presented a short outline of his Sun-centered theory in a published article, *The Commentariolus*, prompting a mixed public reaction. Over the next several years, he lectured on his findings while his colleagues strongly encouraged him to publish the full statement of heliocentricity. Despite this support and Pope Clement VII's informal approval of the theory, Copernicus hesitated to publish it in its entirety because he remained reluctant to fully challenge the accepted view of the universe. Finally, in 1540, after much persuasion by his friends, Copernicus gave permission for the publication of his full and monumental work, *De Revolutionibus Orbium Coelestium—On the Revolutions of the Heavenly Spheres*, commonly called *The Revolutions*. Bedridden with apoplexy and paralysis, Copernicus was handed an advance copy of the book on May 24, 1543, and died later that day.

Divided into six long parts, *The Revolutions* gives a detailed explanation and justification for heliocentricity. It first established that the Earth is spherical, which many still doubted, and discussed the immensity of the heavens. Copernicus then described a system in which the six planets known at that time are centered around the Sun instead of the Earth, with the Earth rotating and revolving about this "visible God." In Copernicus's system, planets' orbits remained in perfect circles on the crystal spheres, not their actual elliptical orbits, because measurements were crude, and true speeds, sizes, and distances in space were totally unknown. Copernicus gave the first logical explanation why each "planet" (derived from the Greek word for wanderer) orbits the Sun in one revolution, and he did so without Ptolemy's numerous and unnecessary epicycles, which didn't even fit the concept of crystal spheres. The Copernican model of the universe also explained why the planets, viewed from Earth, alter their positions while the stars appear to remain fixed in their locations.

The Revolutions helped solve the calendar problem and showed why Ptolemy's theories seemed to be wrong. Yet few people, including astronomers, adopted Copernicus's theory. After all, he failed to explain why there was no great wind from this motion, why a shift of the stars' positions could not be detected, and why all objects (including people) were not hurled off the Earth's surface. Also, his theory conflicted with the Bible. Although the Church did not officially denounce Copernicus when *The Revolutions* was published, there was a negative reaction from many religious leaders. The German theologian Philipp Melanchthon (1497–1560), a strong believer in as-

Figure 1-1

SOLAR SYSTEM—*THE REVOLUTIONS*

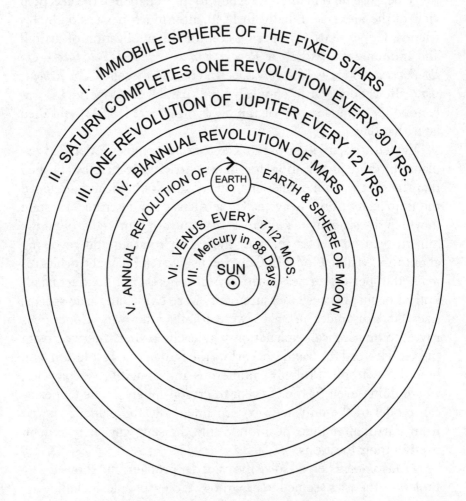

trology and demonology, had tried to prevent publication. The Protestant reformer John Calvin (1509–1564) pointed out that the Bible says the world cannot be moved, and the German preacher and biblical scholar Martin Luther (1483–1546) condemned Copernicus, stating, "The fool will turn the whole science of astronomy upside down." Even the wisest of men could not yet justify dethroning mankind and Earth from the center of the universe to a mere planet of no special significance. Thus, despite Copernicus and *The Revolutions*, Earth remained at the center of the universe for the time being, and it would remain the task of other independent-minded astronomers and mathematicians of the Renaissance to adopt his theory, refine it, and expand it to the level of irrefutability.

Copernicus's proposal of heliocentricity coupled with specific observations of the planets' movements was a bold effort to separate astronomy from philosophy and make it a true science. He rejected Aristotelian/Ptolemaic teachings about why bodies fall to their "natural place," he dethroned mankind and the Earth, and his work eventually led to the revision of mankind's concept of the universe—a fundamental change in thought known as the Copernican Revolution. This profound view also resulted in a new connotation for the term "revolution"—until Copernicus, it meant only the physical movement of heavenly bodies. Perhaps most important, for the first time in history, he introduced the concept that the solar system can be viewed and studied as a structure independent from the stars.

In 1953, in his speech commemorating the 410th anniversary of the death of Copernicus, Albert Einstein said:

> Copernicus not only paved the way to modern astronomy; [he] also helped to bring about a decisive change in man's attitude toward the cosmos. Once it was recognized that the Earth was not the center of the world, but only one of the smaller planets, the illusions of the central significance of man himself became untenable. Hence, Copernicus, through his work and the greatness of his personality, taught man to be modest.

Indeed, as Martin Luther had predicted, Copernicus did turn astronomy upside down. However, he was anything but a fool.

Tycho, the First Great Astronomical Observer, Maps the Universe

A Need for Compasses and Clocks

The Revolutions lay there on the periphery of astronomical thought, like an orphan, benign and unwanted. It was criticized by ecclesiastical leaders but not officially censored by the Church, for the Roman Catholics had not yet begun their fight against heretics and Protestants in full force. A series of events then followed that were to immortalize Copernicus and give heliocentricity a greater significance than even its author had intended. The first was a partial eclipse of the Sun on August 21, 1560, seventeen years after Copernicus's death. It had been predicted by the astronomical tables and had no particular significance in and of itself, except that it was witnessed by Tycho Brahe (1546–1601), then a 13-year-old student at the University of Copenhagen in his native Denmark, and it made such an impression on him that the entire course of his life shifted. He had already begun his studies in rhetoric and philosophy in preparation to follow the tradition of his aristocratic family and become a statesman, but then became fascinated by man's ability to predict events such as this eclipse. He was soon preoccupied with observing the planets, but his family and the Church authorities would not tolerate this unconventional pastime. They forced Tycho to continue his general education and later his law studies at various universities, while he secretly devoted most of his time to his passion for observing the movement of the planets.

His desire to accurately measure the movement of heavenly bodies coincided with Denmark's need for precision in ocean navigation and an ever-increasing demand for accurate compasses and clocks throughout Europe. With this growing acceptance of math, technology, and astronomy, Tycho was finally permitted to devote full time to observations without the earlier disapproval by his family and peers. He completed his long education in 1572 at age 26. Then, on the evening of November 11, 1572, Tycho experienced another astronomical event that would help shape the history of science. He noticed a bright object near the constellation Cassiopeia that had never been there before. It was so unusually bright—brighter than Venus—that it was visible in daylight. In December it began to fade, but remained visible through March 1573. Though there had been stories in history about the appearance of new shining bodies, people had continued to subscribe to the basic Aristotelian doctrine that although the region between the Earth and the Moon could change and decay, the heavens were immutable. In other words, new stars could not be created. In the past, such bright new objects were believed to be comets, since comets were thought to be located only in that area between the Earth and the Moon, that is, *within* the sphere of change and decay, not in violation of Aristotle's teachings. Thus, the critical test of this new heavenly object of 1572 was whether its movement could be detected. If it was moving, it was just another comet. If not, something was rotten in Denmark and in the rest of the universe, as conceived of to date.

Over the several-month period that the object was visible, astronomers throughout Europe applied their crude measuring devices and techniques, and concluded that this new light stood still. Likewise, Tycho observed and measured the phenomenon with his newest sextant, the most sophisticated in existence, and came to the same conclusion—it must be a star, not a comet. Aristotle's principles on the immutability of the heavens were wrong. Nevertheless, the alchemists and astrologers were quick to furnish their own explanations for the nova: It was thought to be a comet condensed from the rising vapors of human sin, composed of poisonous dust drifting down on people's heads; the cause of evil, bad weather, and pestilence.

Thus, the long and awkward journey to the Renaissance continued, as Tycho Brahe complained of the "blind watchers of the sky" and wrote a book describing his detailed observations. Tycho didn't venture to guess *how* the star came into existence—it would take centuries before scientists would determine the explosive process that causes novas. Instead, he concentrated on making accurate observations, and did it so well that in 1576 King Frederick II of Denmark granted him title to the 2,000-acre, 3-mile-long island of Ven, in the sound off Copenhagen, for building and running an observatory at Denmark's expense. Tycho turned Ven into a monument reflecting his grandiose style. Completely funded by the state, Tycho built a large home, the observatory, a chemical laboratory, clocks, sundials, globes, a mill for grinding corn, sixty fish ponds, flower gardens, an arboretum, a paper mill, a printing shop to produce and bind his manuscripts, a windmill and pump for water, and observational inventions and devices that were to become wonders of the world. This extravagant complex, which he christened Uraniborg (meaning heavenly castle), was furnished with a staff of artisans to build the instruments, professional astronomers to record the universe, servants, and all the other workers required to run his little community. Tycho spent the next twenty years at the observatory distinguishing himself as the pioneer of methodical, scientific observation. All with the naked eye, for the telescope had not yet been invented!

Tycho Brahe never fully embraced Copernicus. Indeed, his own concept of the heavens included a modified version of geocentricity, including epicycles similar to Ptolemy's. He was not a creative genius, though he was the first person to apply higher math to observational astronomy when he made the significant discovery that a comet that he saw in 1577 orbited beyond the Moon. His place in history is secured by a much more accurate mapping of the stars and planets and their movements than had been previously done, resulting in the revision of virtually every astronomical table that existed up to that time. He realized that astronomy needs precise data—for the calendar, for navigation, or just for the sake of knowledge. Even Copernicus had failed to recognize the importance of this principle, having relied on the data of Ptolemy, the Greek astronomer Hipparchus, and others, which was largely inaccurate. *The Revolutions* records only twenty-seven observations. In contrast, Tycho undertook tens of thousands. Until Tycho, the emphasis was on qualitative and philosophical theory, not exact measurement. Tycho's work had immediate application for farmers, navigators, and clock makers, and was viewed as revealing God's work and God's plan.

But for Tycho to become an important link in the astronomy/physics chain leading from Copernicus to Isaac Newton, a new and unexpected application of Tycho's wealth of accurate information was first required. It would take a greater mind to somehow seize Tycho's data and discover the greater secrets hidden in his volumes. It would take another step to complete the story that Tycho Brahe began to write unwittingly with his fantastic devices that measured the movement of this Earth, the other planets, and the stars.

Kepler Searches for God's Grand Scheme of the Universe and Discovers the Laws of Planetary Motion

A Mangy Dog . . . Scabs of Chronic Putrid Wounds, and Musical Scales

"Let all keep silent and hark to Tycho," wrote the brilliant and eccentric Johannes Kepler (1571–1630). "For Tycho alone do I wait; he shall explain to me the order and arrangement of the orbits . . . Tycho possesses the best observations . . . He only lacks the architect who would put all this to use." Copernicus had turned his back on Aristotle and Ptolemy and begun to open the door to the future. Through that door was the light of science and knowledge. Tycho proved that Copernicus was justified in standing firm at the threshold, and now Kepler was about to take the first step across it. He was born in Wurttemberg, Germany, the year before Tycho eyed the inspirational nova. He grew up in a poor family, the son of an irresponsible father, but distinguished himself as a good student and was given an education by the state. He left a great gift to humanity, particularly to historians and scientists who came after him, in the form of his prolific writings, most of which still exist. His notes, journals, and books consist of outpourings of all aspects of his life, including detailed chronicles of his childhood and health:

"I was born premature, at thirty-two weeks . . .

1575 . . . I almost died of small pox, was in very ill health, and my hands were badly crippled

1577 . . . On my birthday I lost a tooth, breaking it off with a string which I pulled with my hands

1585–86 . . . During these two years, I suffered continually from skin ailments, often severe sores, often from the scabs of chronic putrid wounds in my feet which healed badly and kept breaking out again. On the middle finger of my right hand I had a worm, on the left, a huge sore

1587 . . . On April 4, I was attacked by a fever

1589 . . . I began to suffer terribly from headaches and a disturbance of my limbs. The mange took hold of me . . . then there was a dry disease

1591 . . . The cold brought on prolonged mange . . ."

"That man has . . . a dog-like nature," wrote Kepler, referring to himself in the third person. "His appearance is that of a little lap dog. His body is agile, wiry, and well-proportioned . . . so greedy that whatever his eyes chanced he grabbed . . . He continually sought the good will of others . . . He is bored with conversation, but greets visitors just like a little dog; yet when the least thing is snatched away from him, he flares up and growls . . . He is malicious and bites people with his sarcasms . . ." His preoccupation with introspection and his maladies was so extreme that it gives the impression of a person who lived more through these mirrorings on paper than in reality. The diary entries were a private, secret ritual, and seemed to be a protective shield against the fundamental unhappiness that originated in his fatherless, troubled, and nomadic childhood. The resulting insecurity spurred him on through a tumultuous and intense life, which he perceived as a never-ending series of challenges, questions, and problems.

Such low self-esteem and high vulnerability were concealed by an aggressive and arrogant outward mien: "This man was born destined to spend much time on difficult tasks from which others shrunk," he said of himself. "In philosophy he read texts of Aristotle . . . He explored various fields of mathematics as if he were the first man to do so . . . He argued with men of every profession for the profit of his mind . . . [and] defended the opinions of Copernicus . . ."

Although Copernicus's *The Revolutions* of a half-century earlier had begun to convince people of a Sun-centered universe, the other ancient astronomical beliefs persisted as the world entered the last decade of the 16th century: *unchanging heavens, perfect circles, celestial harmony, mathematical harmony, and uniform motion.* Kepler would soon play a major role in uprooting those beliefs. Geometry had a particular significance for Kepler because, as he wrote, it "existed before the Creation. It is co-eternal with the mind of God . . . Geometry provided God with a model for the Creation . . . Geometry is God Himself." This combination of the ancient Greek picture of the universe and the principles of basic geometry formed the cornerstone of Kepler's theory.

Early in his life he wrote that he was convinced that his work was destined to reveal great truths about the universe. This self-proclaimed rendezvous with destiny was the subject of much of his voluminous writings. Yet by any objective standard there was no basis for such a prediction. Other astronomers and mathematicians of his time were more intelligent, better educated, and held higher posts, and many certainly had greater opportunities to achieve greatness by virtue of their social status and more likeable personalities. However, on July 9, 1595, while instructing his class in math and astronomy in Graz, Austria, Kepler was struck with the revelation that he'd discovered the secret of the universe. "The delight that I took in my discovery," he later wrote, "I shall never be able to describe in words." His idea was an elaboration of a concept which had developed over the previous centuries, that the universe is built around certain symmetrical geometric figures. He began with two-dimensional shapes. "The triangle is the first figure in geometry," said Kepler. "Immediately I tried to inscribe into the next interval between Jupiter and Mars a square, between Mars and Earth a pentagon, between Earth and Venus a hexagon . . ." He then expanded this theory to the only five solids that can be constructed so that each face of the solid is an identical polygon—the so-called "perfect solids" (cube, tetrahedron, octahedron, dodecahedron, and icosahedron), which he said filled each space between the six planets known at the time (Mercury, Venus, Earth, Mars, Jupiter and Saturn).

With his revelation about geometric harmony, Kepler was more convinced than ever that he'd been correct about his destiny:

Since . . . the full sun illuminated my wonderful speculations, nothing holds me back . . . I dare frankly to confess that I have stolen the golden vessels of the Egyptians to build a tabernacle for my God . . . The die is cast, and I am writing the book to be read either now or by posterity . . . It can wait a century for a reader, as God himself has waited six thousand years for a witness.

From that moment forward, Kepler was obsessed with mathematically proving this concept of the solar system, that is, fitting

Figure 1-2

KEPLER'S GEOMETRIC HARMONY, FROM KEPLER'S *EPITOME OF COPERNICAN ASTRONOMY*

the five perfect solids in the spaces between the six planets. He re-garded it as the ultimate insight to God's plan of the universe. He was convinced there was "a *reason* for the number of planets," and began a frantic pursuit for the truth, sustained by a vision of math-ematical and geometric harmony in the sky. It was not conceivable to him that the fact there are five perfect solids and six planets was merely a coincidence, or that there were actually more than six planets. His tenacity and commitment to his theory sent him on a wild goose chase spanning many years of his life, for his basic the-ory was simply wrong.

Kepler went to work for Tycho in 1600. On his deathbed in 1601, Tycho bequeathed Kepler all of his precise and voluminous records of planetary movements. According to Kepler, Tycho lay there repeating over and over, "Let me not seem to have lived in vain . . . Let me not seem to have lived in vain." Indeed, Tycho's legacy led Kepler to discover three principles that he stated in works published during the period from 1609 to 1618, now known as Kepler's Laws of Planetary Motion:

1. All planets revolve around the Sun in elliptical orbits;
2. A radius line joining any planet to the Sun sweeps out equal areas in equal lengths of time; and
3. The square of the period of revolution of a planet is proportional to the cube of its mean distance from the Sun.

With the third law, he gave a precise timetable of the planetary motions, allowing people to calculate the exact location of each planet at any given time. The second law is depicted in Figure 1-3.

People had spoken and written of an unseen *force* driving the motion of heavenly bodies—commonly conceived of as God. But no one had ever provided such a mechanistic, mathematical interpretation or explanation. In reference to his three laws, Kepler wrote:

Figure 1-3

Kepler's Second Law

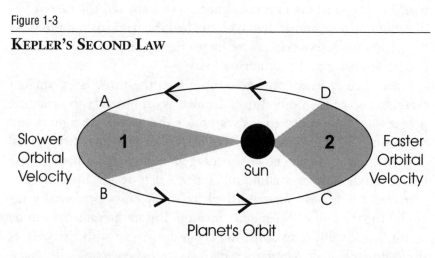

Planet's Orbit

- Time of orbit from A to B equals C to D
- Area 1 equals area 2

Either the *animae motrices* [moving spirit or forces of the planets] are feebler as they are more distant from the Sun or there is only one [force] in the center of all the orbits, that is, the Sun, which impels a body more violently as it is nearer, but which becomes ineffective in the case of the more distant bodies . . .

In the course of trying to understand God's plan in terms of geometric shapes and harmonies, Kepler stumbled across concepts and mathematical principles that accurately describe the relationship between each planet's distance from the Sun and the length of its year. Kepler was the first person in history to tie the planets' motions to their distance from the Sun and to think of cosmological problems in terms of *physical forces* rather than divine guidance. Thus, astronomy and physics became joined for the first time through Kepler's Laws of Planetary Motion. Ironically, these laws proved that his original theories about geometric harmony were not valid. The perfect circular orbits, which Kepler once considered sacred, did not exist. Instead, the orbits were ellipses (first law). The planets did not revolve uniformly. Instead, their velocity was dependent on their distance from the Sun (second law). And there seemed to be this strange relationship, as if the Sun had the power of God. It was devastatingly painful. Indeed, Kepler was actually ashamed of his first law, since it contradicted the sacred belief about circular movements, but he had the wisdom and integrity to express his discoveries as he found them, despite their running counter to his grand geometric theory.

Nevertheless, even after he wrote down the three laws, ancient Greek philosophy loomed over Kepler as an obstacle to scientific progress, and he renewed his search for the geometric figures and perfect orbits, almost as if he could not believe his own discoveries. The magical and subjective still tugged at him to go back to the perfect circles and crystalline spheres of Aristotle and Ptolemy. His *Harmonies of the World*, published in 1618, only incidentally includes the third law of planetary motion. But in the same breath he continued his effort to solve the cosmic mystery with the perfect three-dimensional geometric figures in *combination* with two-dimensional shapes inscribed into the circle. Integral to this grand scheme, Kepler's "harmony" of the universe (in *Harmonies of the World*) is a literal reference to music: "For the extreme diverging intervals of Saturn and Jupiter make slightly more than the octave; and the converging, a mean between the major and minor sixths . . ." The book describes an elaborate mathematical correlation between the planets' movements and musical scales—thus, he envisioned a universe that is a synthesis of ancient Greek philosophy, astrology, astronomy, geometry, and music. Yet the three true laws of planetary motion are hidden within his elaborate conceptualization of the universe and among the false axioms in Kepler's writings. Paradoxically, to the day he died Kepler never recognized the significance of these principles, but seventy years later, Isaac Newton would base an important part of his physics on them.

Though Kepler remained convinced of the "harmonies," his brilliant observations combined with a series of mathematical calculations allowed him to carry Copernicus's baton one more step into the modern world. He did so unwittingly, as he straddled two worlds, torn between the old and new, trying to reconcile two schools of thought.

Over the succeeding decades it became clear that scientists had to rely on facts and reject unsupported speculation when analyzing the physical world; yet, powerful forces persisted in maintaining centuries-old myths and mystic teachings. It was time for courage and thought to coalesce in the hearts and minds of individuals capable of discerning the clear line between astrology and astronomy, between alchemy and chemistry, metaphysics and physics, faith and reason. In Italy at the turn of the 16th century there was a growing and profound effort to place both feet on one side of that line, as the irreconcilability of these views became much more than an intellectual dilemma. It became a matter of life and death.

Chapter Two

The Immovable Earth

"Much of human history can be described as a gradual and sometimes painful liberation from provincialism, the emerging awareness that there is more to the world than was generally believed by our ancestors."

CARL SAGAN
Broca's Brain (1974)

Bruno Is Burned at the Stake as the Church Attempts to Eliminate Threats to Its Authority

A Murder in Rome and a Homecoming in Venice

The chain of events that led to Giordano Bruno (1548–1600) being burned alive at the stake began in 1575 when he secretly read two forbidden commentaries by Desiderius Erasmus (c. 1466–1536), the Dutch humanist and great scholar of the Renaissance. Bruno's impetuousness and his disdain for dogmatic restraint made his fate inevitable, but reading the writings of Erasmus that questioned some Church teachings labeled Bruno a heretic and hastened that fate, as he fought for freedom of thought and expression for the remainder of his life.

In July 1575 Bruno, who'd been ordained as a priest three years earlier, completed the prescribed course in theology at the Dominican Convent of San Dominica Maggiore in Naples. Up to that point, he hadn't publicly criticized the Dominican Order or the Church, but his unorthodox theological views were known, and his enemies reported that he had read those censored works of Erasmus. This was not a matter of Bruno himself writing or advocating disobedience to the Church in those early years of his thinking as a theologian and philosopher. Yet, in the paranoid and tortuous rationale by which the Church attempted to control thought and morality, the mere consideration or discussion of Erasmus's objectionable views was sufficient for a trial for heresy to be prepared against Bruno, forcing him to flee Naples for Rome in February 1576. He was unjustly accused of murder there and also found himself the subject of a second excommunication process. By April 1576, he was again a fugitive, this time settling in Geneva, where he was arrested and excommunicated on the basis of an article he wrote criticizing a Calvinist professor. Nevertheless, Bruno was soon allowed to retract the article—and thus rehabilitated, he was permitted to leave the city.

Between 1580 and 1585, he became a well-known writer in the fields of theology, astronomy, and philosophy, and was a professor in Paris and London, and at Oxford University, where new thought was welcome. During those years, he wrote extensively on the Copernican theory of the solar system and suggested that the universe was infinite. He said the Bible should be followed for its moral teachings but not its astronomical statements. He rejected Aristotelian physics and criticized the Calvinistic principle of salvation by faith alone.

Beginning in 1585, political developments narrowed the margin of tolerance in which Bruno had lived and prospered for the previous five years. Yet, almost as a challenge to those in power in Paris, in 1586 Bruno wrote a series of articles in which he insulted a high government official and renewed his attack on Aristotle. For this he was again forced out, leaving Paris for Germany, where he lectured at various universities and published several articles. He closed out the decade by being excommunicated once again, this time in Helmstedt, Germany, by the local Lutheran Church.

Then, in August 1591, fifteen years after leaving his homeland, he came full circle and returned to Italy, a decision that would ultimately prove fatal, but one in keeping with his tenacious devotion to his principles. When the Venetian patrician Giovanni Mocenigo invited him to return, Bruno viewed it as a safe and reasonable decision. Venice, in 1591, was the most liberal of the Italian states and appeared to be a haven for unorthodox views. Religious tension in Italy and its neighbors had eased after the death of the uncompromising Pope Sixtus V in 1590, and there seemed to be a trend toward religious pacification. Another incentive for returning to Italy was Bruno's hope that he would be appointed to fill the vacancy in the prestigious mathematics chair at the University of Padua. So he began lecturing and writing at Padua in the late summer of 1591, but returned to nearby Venice in early 1592 when it became apparent the chair would be offered to Galileo.

In the spring of 1592, there was a falling out between Bruno and his host Mocenigo, who turned on him unexpectedly and maliciously by denouncing him to the Venetian Inquisition for heresy. In the course of this arrest and trial, Bruno had another opportunity to justify his actions, retract his philosophy, and go on with his life. He admitted errors and pointed out that his theories were philosophy, not theology, and thus were not intended to question the Church's power. Yet, with higher authorities whose memories of Bruno seventeen years earlier were still vivid, the Roman Inquisition demanded his extradition. On January 27, 1593, Bruno became a prisoner of the Holy Office of the Roman palace and began a trial that was to last *seven years*. The charges against him were based primarily on his writings in which he stated that the Earth is not at the center of the universe, the universe is infinite, and the stars are not fixed on a crystal sphere.

The inquisitors informed him that only an unconditional retraction of all his theories would save him. Faced with the choice of uttering lies to indulge his intolerant accusers or maintaining adherence to his principles and losing his life, he declared he had nothing to retract. Pope Clement VIII ordered the death sentence on February 8, 1600, and nine days later, at the Campo di Fiori, Giordano Bruno, bound and his tongue in a gag, was set afire—transforming him from merely a progressive thinker to a martyr for freedom of thought and expression. Bruno's writings later influenced Galileo and became an important source of scientific thought for centuries after his death.

Galileo's Improved Telescope Reveals the Heavens

A Dispute Breaks Out . . . Jupiter's Moons

Born in Pisa, Italy, on February 15, 1564 (two months before Shakespeare's birth), Galileo Galilei was the eldest of seven children and was raised in a household that valued the arts and welcomed new ideas. His father, Vincenzio, was a cloth merchant by trade, and an accomplished musician and composer. In 1574 he moved the family to Florence, where Galileo was sent to the famous Jesuit Monastery school in nearby Vallombrosa when he was 12 years old. He enrolled at the University of Pisa at the age of 17 as a medical student and quickly developed a reputation for obstinately disputing the doctrines handed down from Aristotle and the Greek physician Galen (129–199).

Galileo eventually abandoned his medical studies to pursue mathematics, mechanics, and hydrostatics. His rapid mastery of those subjects can be largely attributed to his mentor and tutor at the University of Pisa, Ostilio Ricci, whose teachings emphasized practical application of math principles, which ideally fit Galileo's developing view of the universe and his prowess as an inventor. In his first year at the university, his attention was drawn to a swinging lamp and the fact that it always seemed to require the same amount of time to complete an oscillation, regardless of the range of the swing. After Galileo verified his observation through experiments, he suggested that the constant regularity of the pendulum could be used to construct a clock that would be highly accurate. He also applied the principle in an invention for measuring the pulse.

Forced to leave school in 1585 for lack of funds, Galileo continued studying on his own and sufficiently developed his knowledge of physics to be appointed a lecturer at the Florentine Academy when he returned to Florence. In 1586, he published an essay describing the invention of the hydrostatic balance (a scale allowing for precise measurements), which brought him to prominence throughout Italy. However, he was still unable to find a steady source of income. But in 1589, at age 25, primarily as a result of a treatise he wrote in 1587 on the center of gravity in solids, Galileo earned the chair of mathematics at the University of Pisa, where he wrote and taught on a variety of subjects, including the laws of motion, for the following two years.

It was at Pisa that Galileo was said to have dropped two balls of unequal weight from the Leaning Tower to prove that they would both fall and accelerate at the same rate, contrary to the writings of Aristotle, who said the heavy ball would reach the ground before the light one. Historians are generally convinced that Galileo never performed that experiment. They surmise that he either conducted a *demonstration* at the Leaning Tower (of a result he already knew) or he performed a "thought experiment" like the following to prove the point: Imagine two unequal balls dropped from the tower at the same time, said Galileo, and also suppose that Aristotle was correct and the heavy ball would fall faster. But now imagine the same experiment with one difference, namely, that the two unequal balls are joined by a string or cable between them. If it were true that the heavy ball moves faster and the light one slower, then the light one will hold back the heavy one. If Aristotle were correct, the two balls tied together would not reach the ground as quickly as the heavy ball alone. But if we assume that the string or cable between the balls has the effect of turning the two balls into a single mass that is *heavier* than either one by itself, the tied balls should drop faster than either one by itself. Whether or not Galileo performed an actual experiment or a thought experiment, his work led to the realization and proof that in a vacuum, where air resistance is not a factor, all objects fall at the same rate. A feather falls as fast as a cannonball.

Galileo was forced to resign from the University of Pisa in the summer of 1592. Historians have various theories concerning his departure. Some maintain that his lectures on his discoveries concerning the motion of falling bodies might have offended and alienated the members of the faculty who still believed in the teachings of Aristotle and who saw their vested interests being threatened by Galileo. Also, he was highly critical of the university's regulations. In another theory, Professor Stillman Drake (translator and biographer of Galileo) suggests that political pressure was used against Galileo by a son of the Grand Duke of Tuscany whom he'd offended. Whatever the reason, Galileo left Pisa and his native Tuscany to fill the mathematics chair at the University of Padua in the Republic of Venice, having been selected over Giordano Bruno (sixteen years Galileo's senior) for the prestigious position. Galileo remained at Padua for eighteen years (1592–1610), teaching math and astronomy, and at the same time accomplishing much of the work that established Galileo's reputation as a scientist and inventor and that was later used by Isaac Newton to form the foundation of modern physics. During the years at Padua, he lived with Marina Gamba and fathered two daughters (in 1600 and 1601) and a son (in 1606).

Galileo's scientific contribution as the pioneer of modern physics was more significant than his work in astronomy or as an inventor. His research and discoveries in physics, which would later be used by Newton, were set forth in these writings:

- 1585–87: Writes essays on motion and centers of gravity of certain objects.
- 1588: Writes landmark paper on motion, *De motu.*
- 1589: Writes essay on logic, entitled *Demonstrations in Science.*
- 1591: Conceives axial rotation of Earth.
- 1593: Composes summary on branch of physics to be known as mechanics.
- 1595: Explains ocean tides in terms of Copernican motion of Earth.
- 1601: Completes *The World System*, and analyzes Kepler's data.
- 1602: Begins studies on magnetism and pendulum motion.
- 1603–04: Develops theorums on motion and falling objects.
- 1612: Publishes book on the motion of objects in water.
- 1623: Writes *The Assayer,* on the scientific method.
- 1624: Begins *Dialogue on the Tides.*
- 1625–31: Further writing on motion, tides, and falling bodies.
- 1632: Publishes *Dialogue on the Two Chief World Systems.*
- 1638: Publishes *Two New Sciences.*

Of all his discoveries and inventions, his most famous is the telescope. Yet, as with the legend of dropping the balls from the Leaning Tower of Pisa, there is also an element of myth in the story about his telescope. Galileo himself wrote, "We are certain the first inventor of the telescope was a simple spectacle-maker who, handling by chance different forms of glasses, looked, also by chance, through two of them, one convex and the other concave, held at different distances from the eye; saw and noted the unexpected results; and thus found the instrument." Credit for the invention is generally given to a Dutch spectacle-maker named Hans Lippershey, who tried to sell the telescope to the government of Holland in 1608 as a military instrument for its war against Spain. After a special committee recommended the purchase, a dispute broke out in Europe among several people, each claiming he was the inventor. Galileo wasn't among them.

By the end of 1608, telescopes were being made and sold throughout Europe for navigation, military uses, astronomy, and novelty items. When the Venetian senate expressed a desire to buy the new seeing device for maritime military uses, a government official aware of Galileo's expertise in instrument-making asked him to make a telescope of his own design. In July 1609, Galileo began experimenting with ways to make the lenses. In just one month he'd made an instrument three times more powerful than any in existence, and presented one to the senate as a gift. By the end of the year he'd made a 30-power telescope, more than tripling his previous efforts. It was this superior talent as an instrument-maker that resulted in Galileo receiving the primary credit for *inventing* the telescope. His decision to quickly turn the device upward to the heavens resulted in his fame as an astronomer and further enhanced the historical link between Galileo and the telescope. That decision also brought about the tumultuous events that would dominate the rest of his life.

"Our sense of sight," Galileo wrote in March 1610 after looking through the telescope, "presents to us four satellites circling about Jupiter, like the Moon about the Earth, while the whole system travels over a mighty orbit about the Sun . . ." This is from his pamphlet *Sidereus Nuncius* ("The Starry Messenger"), which was the first written account of observations of celestial objects through a telescope. In addition to Jupiter's moons, Galileo described the mountainous surface of our own Moon, and explained that the Milky Way consists of stars rather than a white cloudy substance, as previously thought. This work with the telescope immediately led to his appointment as philosopher and mathematician extraordinaire to the Grand Duke of Tuscany. He left his position at the University of Padua to fill this lucrative post and devote more time to research. Praise came from all quarters. He was invited to Rome and honored by a meeting with the Pope on April 1, 1611, where he gave a demonstration of the telescope, and on April 14, 1611, the Academy of Lynxes (a scientific society in Rome) held a banquet in Galileo's honor. For several years, Galileo enjoyed the freedom to report his observations publicly and without concern for how they might conflict with the entrenched beliefs of the day. He had no reason to suspect that these observations would soon become seeds of controversy and tragedy.

Galileo Begins to Publicly Support a Sun-Centered Universe as the Church Issues a Warning

Letters to the Duchess on Sunspots . . . God Stops the Sun for the Israelites

As early as April 4, 1597, Galileo had written to Kepler that he'd "become a convert to the opinions of Copernicus many years ago," yet he continued to teach the Ptolemaic system throughout his eighteen years at Padua because he felt more evidence was needed. But in 1610, after his own observations with the telescope, Copernicanism became a clearly supportable and objective truth for him, and he hinted at these views in his book *The Starry Messenger,* though he stopped short of fully endorsing Copernicanism. His discoveries of the moons of Jupiter, the rings of Saturn, and the phases of Venus revealed the reality of this solar system in a way that Copernicus himself could not have known.

The Church advised Galileo that it disagreed with the interpretations and descriptions in *The Starry Messenger,* but did not interfere with his right to express his views or continue those observations. Then began the conflict that would occupy Galileo's remaining years. In 1613, stepping into the unknown chasm of new ground broken by Bruno, Galileo began to publicly support a heliocentric solar system. This is a new theory, Galileo later said, on which "all my life and being henceforth depends." The flames that engulfed Giordano Bruno in 1600 had since burned out, but the political climate of Italy that had ignited the fire was not very different in 1613. Dancing on the edge of the Church's favor and license, Galileo's teachings, writings, and theories were no longer calmly accepted.

In December 1613, Madame Christina of Lorraine, Grand Duchess of Tuscany (mother of the Grand Duke) had engaged Galileo's friend and student Benedetto Castelli, a mathematician and Benedictine abbot, in a discussion over dinner at the Royal Court. Numerous dignitaries were present, including the Grand Duke, but Galileo wasn't there. Castelli found himself under siege and having to defend Galileo's discoveries against the traditional "wisdom" of the group. After Galileo received a letter from Castelli describing the event, he composed the first of a series of letters that came to be called the *Letters on the Sunspots*, discussing the relationship between science and religion, and sent it to Castelli in the hope that it would be useful if he ever found himself in a similar predicament. Following Bruno's ill-fated lead, Galileo wrote that the Bible should be followed for its moral teachings but does not contain the answers to the mysteries of nature.

During most of 1614, opposition to Galileo and his public views was quietly forming among jealous university colleagues and within the Catholic Church. The first public ecclesiastical attack was launched from the pulpit of Santa Maria Novella in Florence on December 21, 1614, when Father Thomas Caccini denounced Galileo. Caccini first referred to the passage in the Bible when Joshua beseeched God to halt the Sun so the Israelites would have sufficient daylight to sustain their momentum and defeat the Amorites. Based on this, Caccini posed the question: "If God stopped the Sun, how could it be that the Sun wasn't moving around the Earth in the first place?" In a vicious and lengthy condemnation, the priest didn't stop at the Copernican system, but indicted Galileo personally as well as mathematics and all mathematicians as religious and political heretics.

In February 1615, an influential Dominican priest named Niccolo Lorini was given a copy of Galileo's private letter to Castelli, and reported Galileo's "heretical views" to the holy office in Rome. Galileo heard of Lorini's action and, wary of the problems the Castelli letter could cause, prepared a modified version of it and sent it to his friend Piero Dini in Rome and asked him to show it to Cardinal Robert Bellarmine, the Church's chief theologian, along with a cover letter, dated February 16, 1615, downplaying some of the points in the original version that were in conflict with the Scriptures. However, in a letter dated March 12, 1615, Castelli informed Galileo that the Archbishop of Pisa had demanded that Castelli relinquish the original letter to him and that the archbishop had said "it was soon to be made known to you, Galileo, . . . that these ideas are all silly and that they deserve condemnation."

The battle lines were becoming clear. While the opposition coalesced, there was also a loyal and growing band of supporters, including a number of Jesuits. One priest, Paolo Antonio Foscarini, had even written a book devoted to defending the Copernican system from charges that it was inconsistent with the Bible, and had sent a copy to Cardinal Bellarmine for his opinion shortly after the Cardinal had received the revised version of the letter to Castelli. However, Bellarmine resoundingly rejected Foscarini's book and implied grave consequences for those who supported the views of Copernicus and Galileo.

Thus, amid this storm of controversy brewing in the spring of 1615, Galileo was faced with the choice of abandoning his position altogether or demonstrating that the Bible can be reconciled with Copernican theory, for he had never actually believed the two to be inconsistent despite the allegations of his detractors. He decided to defend himself in a reasoned and cautious manner. God's Truth, he wrote, is communicated in two forms—the Bible and Nature. "None of the physical effects that are . . . placed before our eyes . . . should ever . . . be placed in doubt by passages of the Scriptures which seemed to have a different verbal import . . . Two

truths can never contradict one another . . ." The primary vehicle
for supporting his position was an amended and expanded version
of the views originally expressed in Galileo's letter to his friend
Benedetto Castelli. In its new form, it became Galileo's famous *Letter to Madame Christina of Lorraine* (subtitled "Concerning the Use
of Biblical Quotations in Matters of Science"), completed in June
1615.

> Some years ago . . . I discovered in the heavens many things that had
> not been seen before our own age. The novelty of these things, as
> well as some consequences which followed from them in contradic-
> tion to the physical notions commonly held among academic
> philosophers, stirred up against me no small number of professors—
> as if I had placed these in the sky with my own hands in order to
> upset nature and overturn the sciences . . . Showing a greater fond-
> ness for their own opinions than for truth, they sought to deny and
> disprove the new things which, if they had cared to look for them-
> selves, their own senses would have demonstrated to them. To this
> end they hurled various charges and published numerous writings
> filled with vain arguments and they . . . cast against me imputations
> of crimes which must be, and are, more abhorrent to me than death
> itself . . .
>
> I hold the sun to be situated motionless in the center of the revolu-
> tion of the celestial orb while the earth rotates on its axis and re-
> volves about the sun. They . . . resolved to fabricate a shield for their
> fallacies out of the mantle of pretended religion and authority of the
> Bible . . . They had no trouble in finding men who would preach
> the damnability and heresy of the new doctrine from their very
> pulpits . . .

The battle was on for the hearts and minds of all Europeans. But
the aristocracy and Church powers were not ready for a fundamental change in thinking about the universe or, more important,
to relinquish any aspect of authority, including the Church's claim
to be sole interpreter of the Bible. Science as an institution was new
and precarious, particularly physics and astronomy. Religion and
the authority of the Church, on the other hand, were firmly em-

bedded in the minds and culture of the day. In his letter to Madame Christina, Galileo appealed to logic over emotion:

> I do not feel obliged to believe that the same God who has endowed us with senses, reason, and intellect has intended to forego their use and by some other means to give us knowledge which we can attain by them . . . The intention of the Holy Ghost is to teach us how one goes to heaven, not how heaven goes.

Before embarking on a lengthy series of examples showing that the Bible cannot be interpreted literally and that much of it is subject to differing interpretations, Galileo warned of the grave consequences for society and its citizens if no one is allowed to suggest such differing interpretations. "Who indeed will set bounds to human ingenuity?" wrote Galileo. "Who will assert that everything in the universe capable of being perceived is already discovered and known?"

However, it didn't matter whether or not Galileo's reasoning in the letter to Madame Christina was compelling or whether his evidence of the physical movement of the planets was comprehensible by the average person. His effort to rally support for his views was failing because the rise of modern science remained weighted down by the 2,000-year history and system of erroneous beliefs that had been embraced by the vast majority of people. Also, as the Roman Catholic Church continued its fierce attack against the Protestant Reformation that had been so successful during the 1500's, and as there was a broader struggle for authority among the secular forces, the Catholic Church, and the Protestants throughout Europe, Galileo's views gradually became more a symbol for being on one side or the other of a political position than they were a developing area of astronomy and mathematics. In other words, this was more an issue of individual freedom to express views than of the accuracy of the Copernican system. Many of the details of Galileo's proof were far too complex for those who chose not to listen, but even when he invited the Jesuit Fathers to look through the telescope and observe Jupiter's moons themselves, they refused. Instead, they continued to express their dissatisfaction with Galileo's inability to "strictly demonstrate" the truth of his position when it is so clear to every human that the Earth stands still.

Galileo Is Tried for Continuing to Defend Copernicanism

The Decree Is Broken . . . Even So, the Earth Does Move

Against the advice of his friends, Galileo insisted on traveling to Rome in December 1615 to further defend his work. Convinced that Galileo's views could potentially undermine the Catholic Church's fight against Protestantism, on March 5, 1616, Cardinal Bellarmine decreed that Copernicanism is "false and erroneous" and banned the writings of Copernicus, stating that "God fixed the Earth upon its foundation, not to be moved forever." At about the same time, Cardinal Bellarmine warned Galileo not to hold or defend the Copernican doctrine, although it could be discussed as a mere "mathematical supposition." Several years then passed without incident, as Galileo led a quiet life in his house in Bellosguardo near Florence.

However, in 1620 the Church published its "corrections" to Copernicus's *The Revolutions*, and in 1622 the Church created the Institution for the Propagation of the Faith (from which the modern meaning of the word "propaganda" is derived). Despite the warning signs that Galileo must have seen from these developments, he went to Rome in 1624, hoping to secure a revocation of the Decree of 1616. Although unsuccessful in this, he did obtain permission from his friend, the new Pope Urban VIII (a Barberini), to write about "the systems of the world," which Galileo interpreted as permission to publish a fair dialogue examining and comparing the Ptolemaic and Copernican views. Attempting to work within this narrow grant of authority, Galileo returned to Florence and spent the next several years working on his most famous book, *Dialogue on the Two Chief World Systems*. Published in Florence on February 21, 1632, and greeted with praise from scholars across the European continent as a literary and philosophical masterpiece, this book would ultimately destroy Galileo.

Shortly after the book's publication, Galileo's now-estranged friend and former supporter, Pope Urban VIII, established a special papal commission to investigate his writings. Based on the commission's report, Galileo was summoned before the commissary-general of the Holy Office in Rome and tried by the ten judges of the Roman Inquisition for espousing the Copernican theory of the structure of the universe, thereby violating the Decree of 1616. He was not allowed to have a copy of the charges or the evidence, and he had no counsel to defend him.

Galileo, the most celebrated scientist in Europe at that time, placed his faith in the demonstration of his principles and asked the Church officials to suspend the dogma and authority of the Scriptures on which they had based the indictment. But despite his age, his poor health, and, indeed, the clear truth of his spoken and written words, this verdict was reached: Galileo's views, like those of Copernicus, were judged "false and erroneous," and he was ordered to retract his words publicly and never teach Copernican theory again. Thus, thirty-three years after Giordano Bruno had been burned alive, Galileo faced the same choice that had been offered to Bruno. In a decision that some have criticized as damaging the cause of science, Galileo chose life. On June 22, 1633, he gave a lengthy recantation that read in part:

> I, Galileo Galilei . . . aged seventy years . . . and kneeling before you, most Eminent and Reverend Lord Cardinals . . . swear that I have always believed, do now believe, and by God's help will for the future believe, all that is held, preached, and taught by the Holy Catholic and Apostolic Roman Church . . . Whereas, after an injunction had been judicially intimated to me by this Holy Office . . . that I must altogether abandon the false opinion that the Sun is the center of the world and immovable, and that the Earth is not the center of the world, and moves, and that I must not hold, defend or teach . . . the said doctrine . . . I wrote and printed a book in which I discuss this doctrine . . . and for this cause I have been pronounced by the Holy Office to be vehemently suspected of heresy . . . Therefore, desiring to remove . . . this strong suspicion . . . I abjure, curse and detest the aforesaid errors and heresies, and generally every other error and sect whatsoever contrary to the

Holy Church . . . I . . . have . . . sworn, promised and bound myself
as above . . .

Legend has it that as Galileo rose from his knees, he murmured,
"E pur, si muove"—"Even so, it does move." To ensure that he
would comply with the sentence, he was shown the instruments of
torture. He was confined in his villa for the remaining days of his
life under strict house arrest. During those years, he wrote his last
and scientifically most important work, *Two New Sciences.* Because
the Inquisition forbade his printing any book, the manuscript was
smuggled to France for publication. In 1638, Galileo became blind
before he was able to receive a copy of this book. Still a prisoner in
his own house, he died January 9, 1642.

It wasn't until 1757 that the Church removed Galileo's *Dialogue
on the Two Chief World Systems* from its list of banned publications.
On October 31, 1992, 359 years after the Roman Catholic Church
condemned Galileo and forced him to recant, the Vatican formally
acknowledged its error in a statement delivered by Pope John Paul
II to the Pontifical Academy of Sciences. "The theologians who
condemned Galileo," said the Pope, "did not recognize the formal
distinction between the Bible and its interpretation. This led them
unduly to transpose into . . . faith a question which in fact pertained to scientific investigation."

Despite this belated admission, the dispute between Galileo and
the Church still stands as one of history's great symbols of conflict
between reason and dogma, science and faith.

During the one hundred years spanning the lives of Tycho, Kepler,
and Galileo, Aristotle's universe of crystal spheres on circular paths
became a wistful image, Ptolemy's epicycles and geocentric theory
of the universe were disproved, and in the ensuing decades the
Roman Catholic Church lost its control of government and the people. These three individuals were responsible for laying the foundation for the rise of astronomy and the beginning of physics in the
17th century. The scientific method of inquiry and its criterion of
truth, and the modern idea of experimentation, can be attributed

primarily to Galileo. His improvement of the telescope, his resultant observations and writings, his tying theory and experiment to one another, and his use of mathematics in physics all contributed to the remarkable scientific discoveries that were to follow. His life and his work would touch another great figure—later in 1642, the year Galileo died, Isaac Newton was born. Newton would lift the discoveries of Galileo and his other predecessors to an unprecedented height. He would vindicate them all and would incorporate their efforts into the first great scientific discovery in history.

Chapter Three

The Principia

"*It appeared to him reasonable to conclude that this power must extend much farther than was usually thought; why not as high as the moon . . . ? He considered . . . if the moon be retained in her orbit by the force of gravity, no doubt the primary planets are carried round the sun by the like power.*"

HENRY PEMBERTON
A View of Sir Isaac Newton's Philosophy (1728)

"*I do not know what I may appear to the world; but to myself I seem to have been only like a boy playing on the seashore, and diverting myself in now and then finding a smoother pebble or a prettier shell than ordinary, while the great ocean of truth lay all undiscovered before me.*"

ISAAC NEWTON (1727)

The Academy for Experiment was founded in Italy in 1657, and the Royal Society of London was founded in 1660. The first scientific articles were published in England and France a few years later, as Europe gradually passed from old methods and approaches to the modern ways of scientific thought. Thus, Isaac Newton, born in 1642, grew up in the midst of a growing intellectual movement of which he was to become the leading figure at an early age.

Isolated for Eighteen Months, Newton Discovers the Laws of Physics, But Does Not Publish for Over Twenty Years

The Apple and the Moon

Newton was born on Christmas Day near the village of Colsterworth in Lincolnshire County, England, in a small stone farmhouse that still stands in the tiny hamlet of Woolsthorpe. His mother later said the baby was "so little they could put him into a quart pot." Born several weeks premature, he was weak and sickly and wasn't expected to survive, but soon overcame the odds and predictions. His father (also named Isaac) had died three months before, and left the 100-acre farm and the modest two-story house to his wife, Hannah. In 1642, a garden grew to the west of the house and included an apple tree that was to loom large in history.

A month after Newton's third birthday, his mother married Barnabus Smith, a reverend in North Witham, a town a mile from Woolsthorpe. For reasons that young Issac would never understand and are still not clear to historians, his mother moved to Smith's home and left her son in her mother's care. Over the ensuing years Newton and his mother seldom saw each other, as she became preoccupied with raising the three children born to her and Mr. Smith. A few years later, Isaac was old enough to climb trees at Woolsthorpe and see North Witham and the spire of Reverend Smith's church in the distance, reminding him of this traumatic loss for the next several years, until Mr. Smith died and Newton's mother finally returned to Woolsthorpe in 1653, when Newton was 10 years old. During his teenage years, Newton occasionally wrote about this agonizing period. For example, he recalled that he'd felt like "threatening my father and mother Smith to burne them and the house over them." The abandonment also seems to explain his sensitive temperament that was often displayed later in life.

At the age of 12, he enrolled in King's School in Grantham, seven miles from his home, and distinguished himself as an outstanding model-maker, but didn't yet excel at math or science. During these years he was fascinated with forces generated by moving air and rushing water, often visiting windmills and building scale models of them. He also carefully tracked the Sun's movements, constructed sundials and a water clock. Like Leonardo da Vinci and Benjamin Franklin, Newton displayed outstanding mechanical aptitude combined with an innate understanding of geometry, yet he was still a teenager.

At age 18 he enrolled in the famous Trinity College at Cambridge University, where he later met the gifted mathematician Isaac Barrow, who encouraged Newton to study mathematics and optics. Newton received his degree from Cambridge in 1665, but was forced to return to his small home in Woolsthorpe later that year because the Great Plague of London was sweeping through the city. It eventually claimed 75,000 victims, 16 percent of the estimated population of London, causing the university to close for fear of the disease spreading there. This wave of bubonic plague had begun in the late fall of 1664 in a London suburb and reached its peak in the middle of 1665 when it killed over 68,000 people in London that single year. King Charles II and his court fled from London in June, and the British Parliament was forced to move to Oxford for a short session. In 1666, the mortality dropped to 2,000, but Cambridge University did not reopen until the spring of 1667.

During the "plague years"—the eighteen-month period Newton spent at home, beginning when he was only 23 years old—he laid down the foundation of the work that was to revolutionize science. The origin of the famous apple story can be traced to this period. On May 16, 1666, in the course of attempting to conceive of the physical laws that would explain how the Moon revolved around the Earth, Newton happened to be sitting near the apple tree in the garden at Woolsthorpe when he saw an apple drop to the ground. At that moment, he realized that the same central pull of the Earth applied to both objects, the apple and the Moon, and that:

- If the Moon were at rest, like the apple on a tree, it would also fall to the Earth, and
- It is the Earth's pull—though weakened by the great distance between the Moon and Earth—that keeps the Moon from flying out of its orbit.

This led him to develop the law and mathematical formulas that account for the pull of the Earth decreasing as objects are farther away from it. He figured out that such pull weakens inversely as the square of the distance from the center of the Earth. To compare the force at two different distances you would square each of those distances and interchange them. So, for example, if one object (such as a planet) were two times farther from the Sun than another planet of equal mass, the relative force of gravity would be one over two squared, or 1/4. In other words, the Sun and the more distant planet would pull on each other at 1/4 of the force between the Sun and the closer planet. At three times the distance, the force would be one over three squared, or 1/9.

Mass is the *quantity of matter* in a particular object, such as a planet or other heavenly body. The quantity of mass reflects the actual total number of particles (protons and neutrons) in the atoms that make up such planet or other object, as we will see in Chapter 5. Unlike the *weight* of an object, mass doesn't depend on the gravitational pull of the Earth or other body. Therefore, physics uses this universal unchanging measure.

Figure 3-1

NEWTON'S INVERSE LAW OF GRAVITY

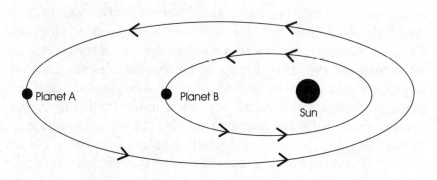

Gravitational force between Planet A and Sun is one-fourth of the gravitational force between Planet B and Sun

It was during this same period at Woolsthorpe that Newton developed the laws of motion:

1. *A body at rest will continue at rest unless a force acts on it, and a body that is moving steadily in a straight line will continue to move with the same speed in a straight line unless a force acts on it.* For example, if you place a ball on a level surface, it won't move without some external force acting on it. If such a force causes it to begin rolling, the ball would continue to roll in that direction *forever* if it weren't for the friction with the surface or an obstacle in its path.
2. *Acceleration (rate of change of motion) is directly proportional to the force.* For example, the more force generated by an automobile engine, the faster the car will accelerate. Twice as much force will double the acceleration.
3. *Every action has an equal and opposite reaction.* For example, the "action" of a bullet being fired by a gun results in the "reaction" of the gun's recoil.

With these laws of motion Newton established the science of mechanics and laid the groundwork for what is now called "classical physics." These principles, including the inverse law of gravity, might seem obvious and simple to today's physicists, but this was a new way of thinking in Newton's time—to find simple, precise mathematical laws from which the observed measurements could be worked out in detail. He was able to rely on Kepler's and Galileo's work to some extent in the development of the inverse law of gravity and the laws of motion, a debt that Newton acknowledged when he said that he "stood on the shoulders of giants . . ." But the principle of universal gravitation and its mathematical formula in particular was a monumental advance beyond Galileo's astronomical discoveries and beyond Kepler's laws of planetary motion, which only told us that the planets move in certain regular orbits related to their distance from the Sun. Although Kepler touched on a "central force," his laws do not provide insight into the dynamics of the solar system or how such a structure as the solar system can function or exist at all. Newton's idea, on the other hand, was that there exists an invisible force that exerts control over matter without having direct physical contact. This was a bold idea in 1666. It demonstrated for the first time the mathematical formulas supporting the concept Copernicus had touched on—how the solar system can be understood as a totally interrelated dynamical structure.

Physically and intellectually isolated during the Woolsthorpe period, Isaac Newton pondered the motions of the heavenly bodies. "I keep the subject constantly before me and wait til the first dawnings open little by little to the full light . . . ," he later wrote. Newton also developed a new branch of mathematics known as the differential and integral calculus and began his work in optics. "All this was in the two plague years of 1665 and 1666, for in those days I was in the prime of my age for invention and minded mathematics and philosophy [meaning science] more than at any time since." In that eighteen months, Newton mastered the basic laws of mechanics, convinced himself they applied to all celestial bodies, discovered the fundamental law of gravity, invented calculus, and made significant advances in the field of optics. Nevertheless, he remained very reluctant to publish any of his discoveries because he felt that further experimentation and proof were needed. Certain calculations didn't seem accurate to Newton, as he'd relied on the generally accepted (but incorrect) value for the diameter of the Earth. The world would have to wait while Newton worked out this and other problems. He became preoccupied with his lengthy correspondence with several other scientists on color, light and optics, telescopes, grinding lenses, and other subjects. He put his notes aside and temporarily turned away from science to engage in other pursuits—charging himself with the safekeeping of these ideas for over twenty years.

Newton Reveals the Laws of the Universe

A Hard Sell in Latin

He returned to Cambridge when it reopened in 1667 and was recognized as a senior fellow of Trinity College. The next year, his mentor Isaac Barrow resigned his position as Lucasian Professor of Mathematics and arranged for Newton, only 25 years old at the time, to succeed him in that prestigious chair (the same position now held by the famous theoretical physicist Stephen Hawking).

By 1684, Newton still hadn't published any of the principles he'd developed at Woolsthorpe. However, in that year he explained his calculations on planetary motion and elliptical orbits to his friend and colleague, the great astronomer and mathematician Edmond Halley (1656–1742), who determined the orbit of Halley's Comet and thus became its namesake. Halley immediately recognized the overwhelming importance of Newton's work and convinced him to publish his discoveries. At first, Newton could not even find his notes. After finally locating them, he began to organize this material into a cohesive body of theories and mathematical demonstrations. Halley solicited financial support from the Royal Society to have this work published, but when the Society backed out, Halley paid the costs of publication himself, negotiated with the printers, edited Newton's writing, wrote the preface and handled all the details of getting Newton's work published. Thus, the world has Halley to thank for the appearance—more than 20 years after Woolsthorpe—of what is now considered the greatest scientific book ever written, *The Mathematical Principles of Natural Philosophy*, commonly known as the *Principia*.

The *Principia* is divided into three "books," although it was published as one volume when it first appeared in 1687. The first book is about the branch of physics called mechanics, and explains the reasons bodies move in the manner they do in empty space. In the second book, Newton deals with the motion of bodies in surroundings that offer resistance, such as air or water. In the third book, using the principles established in the first two books, Newton demonstrates the structure and operation of the entire solar system. He coined the word "gravity" from the Latin word *gravitas*, meaning "heaviness" or "weight." He so accurately explains the motions of the moons of Jupiter, Saturn, and the Earth, as well as the movements of all the planets around the Sun, that few significant improvements were made to his work for 200 years thereafter. He then describes how to calculate the masses of the Sun and planets from the mass of the Earth, which he also demonstrates mathematically. He explains the precise motions of the Moon and that the tides of the seas are due to the gravitational pull of the Moon and the Sun on those waters. He also calculates the Sun's attraction on comets.

The *Principia* is generally considered to be the most prodigious accomplishment of human intellect in the history of mankind, and without question the most important single work ever developed in physics. The book immediately brought great fame to Newton. In 1687, the French scientific periodical *Journal des Savants* (Journal of the Learned) stated: "The work of Mr. Newton is the most perfect treatise on mechanics that can be imagined, it not being possible to provide more precise or more exact demonstrations than those which he gives . . ." A twelve-page review of the *Principia* in the German magazine *Acta Eruditorium* (The Transactions of the Learned) in 1687 called Newton the foremost mathematician of the time. The demand for the *Principia* grew steadily in succeeding years. It was published in several other countries, though usually in the original Latin version. It was revised by Newton for a second edition in 1713 and a third edition in 1726, which was the last edition in his lifetime. The book was translated into English for the first time shortly after Newton's death in 1727, and has since been translated into many other languages.

Despite being recognized as a great work, the *Principia* was so lengthy, complex, and difficult to grasp (particularly in Latin) that it took a full 50 years for the Newtonian scheme of the universe to be widely accepted and taught in schools and universities. Even at Cambridge itself, six years after the *Principia* was published, the professors were still teaching the Cartesian system, based on the French mathematician Descartes's incorrect theories of the solar system developed in the early 1600's. When the second edition of the *Principia* was published in 1713, including new material on the motions of the Moon, comets, and other heavenly bodies, and concerning the resistance of fluids, Newton's associate Roger Cotes still was not able to convince critics that Descartes must be abandoned in favor of Newtonian physics.

Newton Loses Interest in Science and Broadens His Life

The Prophecies of Daniel and Paranoid Delusions

For several years following the publication of the *Principia*, Newton displayed very little interest in scientific matters. The long and tedious process of getting the work published had left him exhausted and indifferent to science, similar to an earlier period when Newton was in his mid-thirties and wrote to a friend that his interest in science was "worn out." In 1689 he was elected to Parliament and served for one year, a position that brought him to London often, where he became friends with the great philosopher John Locke and many men of science who would later play an influential role in Newton's remaining years. During these post-*Principia* years, he wrote extensively about the Scriptures, including correspondence with Locke on the prophecies of Daniel.

After failing to be appointed to several high government posts that he sought, Newton finally was named Warden of the Mint in March 1696. In 1699 he was promoted to Master of the Mint, where he succeeded in resurrecting the seriously debased coinage of the British Empire, and where he received much recognition as an innovative administrator. He retained that position until his death in 1727. At about the time he assumed his position at the Mint, Newton exhibited a renewed interest in science. In 1703 he was elected President of the Royal Society, the most prestigious science organization in Europe at the time, and also held that position until his death. He was the first distinguished man of science to hold that office since the astronomer Christopher Wren (1632–1723) twenty years earlier.

In 1704 Newton published his greatest work on optics, called *Opticks: or a Treatise on the Reflexions, Refractions, Inflexions and Colours of Light*. It included two mathematical treatises that he had written years earlier to support his optics theories. In much the same way that the release of the *Principia* was long delayed, all the notes for *Opticks* were prepared even before Newton came to London in 1689, and as Newton stated in the preface, mostly written in 1675. Primarily because of the publication of *Opticks*, Newton was even more exalted throughout the scientific world. He was knighted in 1705, an honor that had never before been conferred for achievements in science.

Unlike most other great men of science, Newton continued to have long periods when he was not interested in physics, during which he devoted all his time and writing to chemistry, alchemy, his duties at the Mint, and religious matters. Much of his religious writing was collected and published after his death, including a historical account of *Two Notable Corruptions of the Scriptures*, *The Chronology of the Ancient Kingdoms*, and *Observations of the Prophesies of Daniel and the Apocalypse of St. John*. On the subject of chemistry and alchemy, Newton left behind a large library of books. In addition, he engaged in extensive correspondence with his friend, the famous English chemist Robert Boyle, concerning matters of chemical combinations and experiments. In keeping with Newton's secretive personality, he never revealed the purpose or results of his own chemical experiments.

Obviously, even great men are not perfect. Isaac Newton was easily irritated, overly sensitive to the slightest criticism of his work, intent on gaining revenge on his enemies (both real and imagined), and impatient with people of questionable talent or low motivation. Between 1690 and 1692, after he'd served in Parliament and when his interest in science was at one of its low ebbs, he repeatedly accused his close associates and friends John Locke and Charles Montagu of deceiving him and attempting to harm his reputation. He recovered from these paranoid delusions by 1693. He seemed to relish his public disputes with other scientists, such as the British astronomer John Flamsteed, who challenged Newton's work in optics and astronomy, and the German mathematician Gottfried Wilhelm Leibniz, whom Newton accused of plagiarizing his own work in optics and the development of differential calculus. In the first year of his presidency of the Royal Society, Newton wrote, "Our Society decays and produces nothing remarkable . . . 'tis governed by persons that either value nothing but their own interests, or understand little but vegetables . . ." (referring to the naturalist Sir Hans Sloane, the Secretary of the Royal Society). These disputes and others flamed throughout Newton's life and were either the cause or the effect of his embitterment.

Although this book doesn't attempt to rank the seven greatest scientific discoveries in history or their discoverers in any particular order, it's appropriate that Isaac Newton comes first. Newton was not only first chronologically (with five of the remaining six great discoveries taking place in the 20th century), but much of what was discovered after him would not have been possible without his work. The impact of Newton's discoveries continues in the 20th century in many fields of science. The wave theory of light uses Newton's laws of motion, as does the kinetic theory of heat. Newtonian theory was important in the development of our understanding of electricity and magnetism, and in the discoveries of electrodynamics and optics by Faraday and Maxwell. Newton's physics guided science for over 200 years—until the early 20th century, when Einstein demonstrated that physics needed to grow beyond the Newtonian framework. Because Isaac Newton supplied the basic principles on which much of science has been built and led the transformation that brought modern science into existence, he will remain the supreme symbol of the power of the human mind to apply reason and the scientific method to the manner in which the physical universe operates.

Part Two

The Structure of the Atom

In Focus

Two hundred fifty million hydrogen atoms packed together side by side would be about one inch long. If an atom were blown up to the size of the Superdome in New Orleans, the nucleus (made of protons and neutrons) would be the size of a pearl in the center of the structure, with the electrons orbiting at thousands of miles per second at the outer perimeter of the building.

The nucleus is extremely heavy, accounting for 99.95 percent of the total weight of the atom, with each proton having a mass about 1,836 times greater than each electron. That is, these electrical charges *weigh* something! However, the "parts" or subatomic particles that make up each atom and each element (hydrogen, oxygen, gold, silver, aluminum, uranium, and all the other chemical elements) are colorless, tasteless, odorless, and without texture. The subatomic particles are not hard, soft, dull, or shiny even though the ninety-two naturally occurring elements make up everything that exists on Earth—oceans, rocks, air, all plants and animals, including us. The *firmness* and other characteristics of these everyday objects made of this "stuff" results from the interrelationship of electrical forces of and between subatomic particles.

Matter in its most fundamental form consists solely of these electrical charges. This substance from "nothing" that results when electrical charges combine to form atoms and when atoms combine to form elements and molecules, would qualify as the greatest illusion there is—except it is reality.

Part Two of this book discusses how scientists discovered this staggering vision of the complexity of the physical world, identifies those primarily responsible for that dramatic discovery, and explores the incredible story of how we came to unearth the key that released the enormous energy locked within the atomic nucleus.

Chapter Four

Good Chemistry

"First we must inquire whether the elements [earth, air, fire and water] are eternal or subject to generation and destruction. . . . It is impossible that the elements should be generated from some kind of body. That would involve a body distinct from the elements and prior to them."

ARISTOTLE
On the Heavens (c. 330 B.C.)

In *The Ascent of Man,* Jacob Bronowski wrote, "The genius of people like Newton and Einstein is that they ask simple, innocent questions which turn out to have answers with enormous effects on society and individuals." Asking such questions is an ancient tradition begun in Greece, which was first inhabited 8,000 years ago and was the first enduring civilization in Europe. Aristotle coined the term "physics" from the Greek word "physis" meaning nature, to designate the study of nature. Most of the teachings and theories of Socrates, Plato, and Aristotle were proved inaccurate and were abandoned during the Renaissance. Yet modern science can still trace its roots back more than two thousand years to ancient Greece—not because these or other ancient Greek philosophers made any scientific discoveries, but because they believed the physical world and universe could be understood through rational thought. This reliance on logic and reason ultimately grew into the scientific method of modern science that was applied so well by Newton. Beginning in the same era as Newton's great work, other scientists were attempting to understand the composition of matter at its most fundamental level. This exploration would take centuries to reach fruition after the 17th century, and actually had its origin in the "physis" of ancient Greece.

Modern Atomic Theory Traces Its Roots to Ancient Greece

Sound and Souls Made of Atoms

In about 500 B.C., the Greek philosopher Anaxagoras inquired into what matter is made of, and imagined cutting in half any bit of matter—say, a ball of silver—and then halving it again and again and again. After contemplating this, he proposed his theory that a person could continue halving any piece of matter without ever reaching an end. In other words, all matter is infinitely divisible.

On the northern shore of the Aegean Sea was the Greek town Abdera (now Avdhira), founded by refugees from the Persian invasion of Lydia in the year 540 B.C. A man named Leucippus, who settled there in 478 B.C., had a theory about matter that was fundamentally different from Anaxagoras's conclusions and from anything else espoused prior to that time. Only fragments of his writings remain, so we have to rely on the writings of his pupil, Democritus, who popularized Leucippus's thoughts, and on Aristotle's characterization of Leucippus's philosophies, which are often referred to by Aristotle in his voluminous writings. Leucippus's theory stated that two things and only two things exist in the universe: atoms and the void. Leucippus believed that everything we see is made of atoms (in Greek "átomos," meaning not cuttable) that move through the void of empty space. He conjured up these hypothetical bits of matter and said they're too small to be seen; they're solid, so solid that they cannot be divided in any way (stable, unchangeable and indestructible); they occur in various geometric forms (which explains their capacity to combine to form all the things in the world); *and they are perpetually in motion.* They form the material that we see, through their collisions and re-groupings. He couldn't imagine that this magnificent universe is made from formless nothingness with no basic identity. Otherwise, how could material things retain their individuality, and how could living things pass along their complex images to their offspring without some type of indestructible pieces of matter?

Leucippus's atomic theory was later used by other philosophers to explain a wide variety of phenomena. For example, sound was supposedly generated by atoms of sound hitting atoms of air, which then struck the ear. Even the human soul was composed of atoms. As with almost all the enduring principles handed down by ancient Greece, Leucippus's atomic theory and others' application of it was totally based on speculation. It was philosophy, not science. But to Leucippus's credit, he proposed a concept that resurfaced 2,400 years later in the modern understanding of the structure of the atom.

Others in ancient Greece pondered the nature of matter before and after Leucippus and Anaxagoras. Empedocles (c. 495–435 B.C.) was the first to propose that earth, air, fire, and water were the elemental substances, while a host of other Greek philosophers attempted to understand and explain our physical surroundings. Thales of Miletus (c. 625–547 B.C.) proposed that water was the basis of all matter. Anaximander of Miletus (c. 610–545 B.C.) spoke vaguely of a "primary substance" that had the capacity of all types of matter and their properties (hot, cold, wet, dry, etc.). Anaximines of Miletus (c. 585–524 B.C.) proposed that air is the primary substance and transforms into other materials. Plato and Aristotle adopted Empedocles's four elements, and added their personal spin on how matter and the physical universe are put together. Though people adhered to the belief in the four elements through the Middle Ages, Leucippus's atomic theory would ultimately succeed that belief.

Procedures and Apparatus of Medieval Alchemy Lay Foundation for Modern Chemistry

Yet Earth, Air, Fire and Water Persist

Beginning thousands of years ago, people valued gold as a rare and beautiful substance, and also recognized that gold had a unique ability to resist decay and corrosion. Since there was no known acid or other substance that could damage gold, it was thought to have a quality of permanence that could be transmitted to humans. Therefore, every medicine for fighting aging contained gold as an essential ingredient, and doctors urged people to drink from gold cups to prolong life.

This universal desire for gold spawned alchemy as a formal discipline in the 1st century A.D. It first arose among Greek scholars, then spread to eastern Mediterranean countries, and finally to Spain and Italy in the 12th century. Though the attempt to produce gold from other substances was the original and central purpose of alchemy, a number of physician-alchemists in Europe in the Middle Ages attempted to produce medicines that were not dependent on gold or related to it. In an effort to produce medicinal essences and spirits from raw materials, such as herbs, this group of alchemists improved methods of separating elements by distillation. For example, as early as the 13th century, Thaddeus of Florence identified the medical benefits of alcohol distillates taken internally and applied topically. Paracelsus (1493–1541), the German-Swiss physician and alchemist, was the first person to unite medicine with chemistry through his use of remedies that contained mercury, sulfur, iron, and copper sulfate. This led to the use of steam distillation and improved equipment.

The development of apparatus and the extensive efforts to break down or distill substances laid the foundation for modern chemistry, but as true science began to evolve during the Renaissance, the study of alchemy *impeded* the birth of modern chemistry. Even as Francis Bacon's teachings on the scientific method led people away from Greek philosophy toward reliance on empirical evidence (that is, what can actually be observed and/or measured), the four elements (earth, air, fire, and water) lived on, and there was no recognition that these four substances are made up of a *combination* of basic elements.

Greek Philosophy Gives Way to the Scientific Method as Elements Are Discovered

Phlogiston and Guillotines

Chemistry—the study of the laws that govern the behavior of the elements—finally broke free from alchemy in 1661 when the Irish chemist Robert Boyle (1627–1691) published *The Skeptical Chymist*, in which he demonstrated that there was no basis for regarding earth, air, fire and water as elemental. Boyle, one of the founding members of the Royal Society of London in 1660, laid the foundation for qualitative analysis using flame colors, spot tests, precipitates, and other analytical tools in an effort to figure out the basic elements comprising minerals and other materials. In 1662, he determined that the volume of gas is inversely proportional to the pressure on it. So, for example, doubling the pressure on a particular volume of gaseous substance causes it to become half that volume, and reducing the pressure to half of the original pressure causes it to double in volume. This became known as Boyle's Law and led to his thesis that gases must be made up of tiny "corpuscles" with a large amount of empty space between them, which is the reason they can be squeezed into smaller volumes. Isaac Newton concurred with Boyle's corpuscle explanation when he referred to atoms as "the small particles of bodies with certain powers, virtues, or forces" by which they could act on each other. With this concept, Boyle declared that Leucippus must have been correct—that these corpuscles are comparable to Leucippus's atoms. Having resurrected Leucippus's version of indivisible particles, Boyle became one of the first significant figures in science to turn his back on Anaxagoras's and Aristotle's teachings about infinitely divisible matter and to shatter the persistent belief that earth, air, fire, and water are the fundamental materials of the universe.

Joseph Black (1728–1799), born in France and educated in Scotland, began experimenting on the constituents of air in the 1750's, which led to his discovery of carbon dioxide and nitrogen. Beginning with Black, scores of other European chemists were attracted by the curious and unique properties of gases and began to focus on their makeup. These studies led to the realization that there is a group of gaseous substances unique from one another that *combine* to make up "air." Therefore, the concept of air as *one element* gave way to "gas" being considered a *state of matter*, along with liquid and solid states. This developing concept rapidly led to the further realization that combinations of several fundamental substances make up all material on Earth.

In 1729 Georg Ernst Stahl, physician to the King of Prussia, had "invented" the concept of phlogiston, a mysterious, invisible, colorless, odorless, and tasteless substance that supposedly had "negative weight." Stahl proposed phlogiston to explain why certain substances became heavier when burned, and as an excuse to cover up the increasing failings of Aristotelian theories to which many people still clung. Supported by a number of leading chemists, the phlogiston theory persisted for decades. For example, the British chemist Joseph Priestley (1733–1804) was the first to isolate oxygen (in 1774) and describe its role in combustion and respiration, but he called it "dephlogisticated air" and didn't understand the full significance of this discovery.

Antoine Laurent Lavoisier (1743–1794), the son of a wealthy Parisian lawyer, pursued the family tradition and received his license to practice law in 1764. But within two years he was drawn back to his desire to learn more about science, an interest first experienced during his earlier education in math, astronomy, chemistry, and botany. By 1772 he'd disproven several of the ancient Greek principles about earth, air, fire, and water, and developed a reputation for exact quantitative procedures and brilliant experiments. Lavoisier possessed the insight and vision necessary to get over the phlogiston hump through his careful measurement of the weight of substances to determine the effect of heating them, and his development of the principle that *a substance could be considered elemental only if it failed to break down into simpler substances when treated chemically.* He expanded the list of known elements to thirty-three, although some were erroneous. From 1776 to 1782, Lavoisier conducted experiments in which he isolated oxygen in air and furthered Priestley's work on oxygen's role in combustion and respiration. The term "oxygine" (derived from Greek for acid former) was used for the first time in a memoir by Lavoisier, dated September 5, 1777. In a 1783 paper (succinctly titled "On the Nature of Water and on Experiments that Appear to Prove That This Substance Is Not Properly Speaking an Element, but Can Be Decomposed and Recombined") Lavoisier reported to the French Academy of Sciences that water was the product of combining hydrogen and oxygen. In a subsequent paper delivered to the Academy, Lavoisier presented a logical analysis about the substance that we now call oxygen. Through Lavoisier's sensitive balance instrument, keen insight, and inductive reasoning, he completed Boyle's effort to vanquish the Greek concept of earth, air, fire, and water, once and for all, and he put the tenacious phlogiston theory to rest. For this and other work, he's now considered the father of modern chemistry.

Lavoisier had been active in political affairs his entire adult life, and devoted much of his career to public service, including positions in the French government from 1768 to 1790 in the areas of economics, agriculture, education and social welfare. In the aftermath of the French Revolution of 1789, despite his many contributions as a reformer and political liberal and despite his participation in the Revolution, he came under attack because of his status as a wealthy member of the French aristocracy, but primarily because of a position he'd held in 1768 in the Ferme-Générale, the country's tax collecting agency. When the Reign of Terror commenced in 1793, resulting in the suppression of the French Academy of Sciences and other learned societies, Lavoisier was arrested. On May 8, 1794, after a one-day trial, the prospect of Lavoisier's further contributions to science and rational thought were prematurely halted at age 50 at *Place de la Concorde* in Paris, as his great mind fell into the blood-soaked basket at the foot of the guillotine along with twenty-seven other former members of the Ferme-Générale.

Dalton Develops the First Modern Theory of Atoms and Mendeleev Constructs a Conceptual Framework for the Elements

Ultimate Particles and Gaps in the Table

It was then left to the son of a poor Quaker weaver in rural England to establish the idea of chemical atomism. In October 1803, John Dalton (1766–1844), professor of mathematics and physical sciences at New College, Manchester, delivered a paper to the Manchester Literary and Philosophical Society concerning the mixture and solubility of different gases: "The circumstance depends upon the weight and number of the ultimate particles of several gases . . . An enquiry into the relative weights of the ultimate particles of bodies is a subject, as far as I know, entirely new. I have lately been prosecuting the enquiry with remarkable success." One month before this speech he'd prepared a table of comparable atomic weights, based largely on calculations made by Lavoisier and others.

Table 4-1

DALTON'S 1803 TABLE OF THE RELATIVE WEIGHTS OF THE ULTIMATE PARTICLES

Hydrogen	1.0	Nitrous oxide	13.7
Azot	4.2	Sulphur	14.4
Carbone	4.3	Nitric acid	15.2
Ammonia	5.2	Sulphuretted hydrogen	15.4
Oxygen	5.5	Carbonic acid	15.3
Water	6.5	Alcohol	15.3
Phosphorous	7.2	Sulphureous acid	19.9
Phosphuretted hydrogen	8.2	Sulphuric acid	25.4
Nitrous gas	9.3	Carburetted hydrogen	6.3
Ether	9.6	Olefiant gas	5.3
Gaseous oxide of carbone	9.8		

We had come full circle from Leucippus in Greece in 478 B.C. to John Dalton in England in the first decade of the 1800's. Now there was a theory—more refined than Boyle's corpuscle theory—derived from two centuries of chemical experiments in which Dalton reached these conclusions:

- The elements are made up of tiny indivisible particles, which Dalton called "atoms" in honor of Leucippus and Democritus.
- The atoms of each element are all alike but differ from the atoms of every other element.
- Chemical combination occurs when the atoms of two or more elements form a "firm union."

It hadn't been necessary for Kepler or Newton to understand the atom to figure out the laws of gravity or motion. But people searching for the ultimate structure of matter in the early 1800's began to realize that they were on the path to understanding and unlocking this secret. The Swedish physician and chemist Jons Jakob Berzelius (1779–1848) began his research in analytical chemistry soon after he became aware of Dalton's atomic theory and was one of the first scientists to recognize its great significance. Beginning in 1810, Berzelius made major contributions in

Table 4-2

DISCOVERY OF THE ELEMENTS

Before 1700	*1700–1799*	*1800–1849*	*1850–1899*
Antimony	Beryllium	Aluminum	Actinium
Arsenic	Chlorine	Barium	Argon
Bismuth	Chromium	Boron	Cesium
Carbon	Cobalt	Bromine	Dysprosium
Copper	Fluorine	Cadmium	Gadolinium
Gold	Hydrogen	Calcium	Gallium
Iron	Manganese	Cerium	Germanium
Lead	Molybdenum	Erbium	Helium
Mercury	Nickel	Iodine	Holmium
Phosphorus	Nitrogen	Iridium	Indium
Silver	Oxygen	Lanthanum	Krypton
Sulphur	Platinum	Lithium	Neodymium
Tin	Strontium	Magnesium	Neon
Zinc	Tellurium	Niobium	Polonium
	Titanium	Osmium	Praseodymium
	Tungsten	Palladium	Radium
	Uranium	Potassium	Rubidium
	Yttrium	Rhodium	Samarium
	Zirconium	Ruthenium	Scandium
		Selenium	Thallium
		Silicon	Thulium
		Sodium	Xenon
		Tantalum	Ytterbium
		Terbium	
		Thorium	
		Vanadium	

determining accurate atomic weights and created the standard system of chemical symbols and chemical proportions still in use today. Over the next few decades, Dalton's atomic theory was gradually accepted by other scientists, and a greater understanding of "firm unions," or molecules, was realized through the work of the Italian physicist Amedeo Avogadro (1776–1856) and the Italian chemist Stanislao Cannizzaro (1826–1910). Their combined theories provided a means to determine molecular weight and size. Introduced in 1811, Avogadro's two-part hypothesis held that: *(1) the ultimate particles are not necessarily atoms, but may be groups of atoms joined to form molecules; and (2) equal volumes of gases contain equal numbers of molecules.* Despite its accuracy and usefulness, Avogadro's hypothesis was not consistent with the accepted principles of that time. Therefore, it went largely unnoticed until Cannizzaro developed additional evidence to support it and presented the hypothesis to a chemists' congress in Germany in 1858. Following this, chemistry rapidly matured into a true science, and the elements were organized into a new conceptual framework and system.

Table 4-3

REPLICA OF MENDELEEV'S VERTICAL TABLE OF 1869

			K=39	Rb=85	Cs=133	—	—
			Ca=40	Sr=87	Ba=137	—	—
			—	?Yt=88?	?Di=138?	Er=178?	—
			Ti=48?	Zr=90	Ce=140?	?La=180?	Th=231
			V=51	Nb=94	—	Ta=182	—
			Cr=52	Mo=96	—	W=184	U=240
			Mn=55	—	—	—	—
Typische			Fe=56	Ru=140	—	Os=195?	—
Elements			Co=58	Rb=104	—	Ir=197	—
			Ni=59	Pd=105	—	Pt=198?	—
H=1	Li=7	Na=23	Cu=63	Ag=108	—	Au=199?	—
	Be=9,4	Mg=24	Zn=66	Cd=112	—	Hg=200	—
	B=11	Al=27,3	—	In=113	—	Tl=204	—
	C=13	Si=28	—	Sn=118	—	Pb=207	—
	N=14	P=31	AS=75	Sb=122	—	Bi=208	—
	O=16	S=32	Se=78	Te=125?	—	—	—
	F=19	Cl=35,5	Br=80	I=127	—	—	—

By 1869 sixty-three elements were known. Dmitri Mendeleev (1834–1907), born in Siberia, Russia, the youngest of seventeen children of a school teacher, prepared a set of cards in which the properties (atomic weight, specific gravity, volume, valence, specific heat, etc.) of the elements was tabulated on a separate card. By arranging and rearranging the cards, Mendeleev figured out that such properties recurred "periodically" when the cards were arranged on the basis of increasing atomic weight. He then prepared a table to reflect this principle, which was published in 1869, as shown in Table 4-3.

Mendeleev developed principles that were applied in the table, including listing according to increasing weight, and placing the symbol for hydrogen away from the elements that followed hydrogen in weight. Because he had the insight to see that many elements had not yet been discovered, he left open spaces in the periodic table—like a jigsaw puzzle, he imagined the missing pieces, based on the "color and shape" of those that were there. For example, he predicted that an unknown element with the atomic weight of 44 would be found for the space following calcium. This prediction was proven correct in 1879 when the Swedish chemist Lars Fredrik Nilson (1840–1899) discovered scandium.

Mendeleev's table developed into the modern periodic table, shown in Table 4-4, one of the most important tools in chemistry. The vertical columns of the modern periodic table are called *groups* and the horizontal rows are called *periods*. The atomic number of an element is the number of protons in the nucleus of the atoms of that element. Neutrons are ignored. Not only does the modern periodic table succinctly organize all the elements, it clearly illustrates that they form "families" in rational groups or trends, based on their salient characteristics (atomic weight, diameters, density, and energy).

Table 4-5 lists all the naturally occurring elements in order of atomic number. It also gives the atomic weight, which is based on a scale that uses a specific type of carbon atom (carbon-12) as the standard against which all other elements' weights are valued or measured. Carbon-12 is an "isotope" of the carbon atom, meaning it is one "species" of carbon among others of nearly identical chemical behavior but with different atomic masses. The majority of elements are mixtures of several isotopes—that is, the ratio of neutrons to protons varies slightly. The atomic weights of elements with no "stable" (i.e., non-decaying) isotope are stated in brackets on Table 4-5 as the mass of their most stable isotope.

Table 4-4

THE MODERN PERIODIC TABLE

1 H																	2 He
3 Li	4 Be											5 B	6 C	7 N	8 O	9 F	10 Ne
11 Ne	12 Mg											13 Al	14 Si	15 P	16 S	17 Cl	18 At
19 K	20 Ca	21 Sc	22 Ti	23 V	24 Cr	25 Mn	26 Fe	27 Co	28 Ni	29 Cu	30 Zn	31 Ga	32 Ge	33 As	34 Se	35 Br	36 Kr
37 Rb	38 Sr	39 Y	40 Zr	41 Nb	42 Mo	43 Tc	44 Ru	45 Rh	46 Pd	47 Ag	48 Cd	49 In	50 Sn	51 Sb	52 Te	53 I	54 Xe
55 Cs	56 Ba	57 La	72 Hf	73 Ta	74 W	75 Re	76 Os	77 Ir	78 Pt	79 Au	80 Hg	81 Tl	82 Pb	83 Bi	84 Po	85 At	86 Rn
87 Fr	88 Ra	89 Ac															

58 Ce	59 Pr	60 Nd	61 Pm	62 Sm	63 Eu	64 Gd	65 Tb	66 Dy	67 Ho	68 Er	69 Tm	70 Yb	71 Lu
90 Th	91 Pa	92 U											

Table 4-5

LIST OF NATURALLY OCCURRING ELEMENTS

Atomic No.	Name	Symbol	Weight	Atomic No.	Name	Symbol	Weight
1	Hydrogen	H	1.00794(7)	42	Molybdenum	Mo	95.94(1)
2	Helium	He	4.002602(2)	43	Technetium	Tc	[98]
3	Lithium	Li	6.941(2)	44	Ruthenium	Ru	101.07(2)
4	Beryllium	Be	9.012182(3)	45	Rhodium	Rh	102.90550(3)
5	Boron	B	10.811(5)	46	Palladium	Pd	106.42(1)
6	Carbon	C	12.011(1)	47	Silver	Ag	107.8682(2)
7	Nitrogen	N	14.00674(7)	48	Cadmium	Cd	112.411(8)
8	Oxygen	O	15.9994(3)	49	Indium	In	114.818(3)
9	Fluorine	F	18.9984032(9)	50	Tin	Sn	118.710(7)
10	Neon	Ne	20.1797(6)	51	Antimony	Sb	121.760(1)
11	Sodium	Na	22.989768(6)	52	Tellurium	Te	127.60(3)
12	Magnesium	Mg	24.3050(6)	53	Iodine	I	126.90447(3)
13	Aluminum	Al	26.981539(5)	54	Xenon	Xe	131.29(2)
14	Silicon	Si	28.0855(3)	55	Cesium	Cs	132.90543(5)
15	Phosphorous	P	30.973762(4)	56	Barium	Ba	137.327(7)
16	Sulfur	S	32.066(6)	57	Lanthanum	La	138.9055(2)
17	Chlorine	Cl	35.4527(9)	58	Cerium	Ce	140.115(4)
18	Argon	Ar	39.948(1)	59	Praseodymium	Pr	140.90765(3)
19	Potassium	K	39.0983(1)	60	Neodymium	Nd	144.24(3)
20	Calcium	Ca	40.078(4)	61	Promethium	Pm	[145]
21	Scandium	Sc	44.955910(9)	62	Samarium	Sm	150.36(3)
22	Titanium	Ti	47.867(1)	63	Europium	Eu	151.965(9)
23	Vanadium	V	50.9415(1)	64	Gadolinium	Gd	157.25(3)
24	Chromium	Cr	51.9961(6)	65	Terbium	Tb	158.92534(3)
25	Manganese	Mn	54.93805(1)	66	Dysprosium	Dy	162.50(3)
26	Iron	Fe	55.845(2)	67	Holmium	Ho	164.93032(3)
27	Cobalt	Co	58.93320(1)	68	Erbium	Er	167.26(3)
28	Nickel	Ni	58.6934(2)	69	Thulium	Tm	168.9342(3)
29	Copper	Cu	63.546(3)	70	Ytterbium	Yb	173.04(3)
30	Zinc	Zn	65.39(2)	71	Lutetium	Lu	174.967(1)
31	Gallium	Ga	69.723(1)	72	Hafnium	Hf	178.49(2)
32	Germanium	Ge	72.61(2)	73	Tantalum	Ta	180.9479(1)
33	Arsenic	As	74.92159(2)	74	Tungsten	W	183.84(1)
34	Selenium	Se	78.96(3)	75	Rhenium	Re	186.207(1)
35	Bromine	Br	79.904(1)	76	Osmium	Os	190.23(3)
36	Krypton	Kr	83.80(1)	77	Iridium	Ir	192.217(3)
37	Rubidium	Rb	85.4678(3)	78	Platinum	Pt	195.08(3)
38	Strontium	Sr	87.62(1)	79	Gold	Au	196.96654(3)
39	Yttrium	Y	88.90585(2)	80	Mercury	Hg	200.59(2)
40	Zirconium	Zr	91.224(2)	81	Thallium	Tl	204.3833(2)
41	Niobium	Nb	92.90638(2)	82	Lead	Pb	207.2(1)

Table 4-5 (Continued)

LIST OF NATURALLY OCCURRING ELEMENTS

Atomic No.	Name	Symbol	Weight	Atomic No.	Name	Symbol	Weight
83	Bismuth	Bi	208.98037(3)	88	Radium	Ra	[226]
84	Polonium	Po	[209]	89	Actinium	Ac	[227]
85	Astatine	At	[210]	90	Thorium	Th	232.0381(1)
86	Radon	Rn	[222]	91	Protactinium	Pa	231.03588(2)
87	Francium	Fr	[223]	92	Uranium	U	238.0289(1)

Scientists soon realized that to fully understand the physical basis for the laws of chemistry, they would have to determine the actual behavior of the atoms, which was considered impossible until late in the 19th century. Also, for most physicists in the late 1800's, Dalton's description of the atom fell short because it didn't seem to fit into the Newtonian system of physical principles and didn't explain the electrical behavior of matter. One line of inquiry regarding such electrical behavior was physicists' efforts to better understand the behavior of gases when combined with electricity. Numerous experiments on this phenomenon were conducted by physicists and chemists throughout the 19th century. In 1879 the British chemist and physicist William Crookes (1832–1919) reported the results of his extensive research on electrical discharges through gas. In his experiments, he used a vacuum tube of his own design (now known as a Crookes tube), which included a cathode (that is, a negative electrode) and an anode (positive electrode) to allow a charge to pass through the tube. Crookes noted that rays of unknown origin and composition were coming from the tube when electricity was passed through the gas in the tube. He was able to determine several characteristics of the rays, such as that the rays travel in straight lines and produce phosphorescence and heat in certain materials. Over the next fifteen years, a number of other chemists and physicists experimented with cathode rays and Crookes tubes in an effort to understand the composition and characteristics of the mysterious rays.

As scientists approached the 20th century, there was much greater motivation for discovering and understanding what matter is made of than there had been in 478 B.C. when Leucippus first put forth his crude atomic theory. When it became apparent that all atoms shared certain characteristics and that the atom might even be composed of smaller components, physicists and chemists seized the baton from their predecessors in the last decade of the 19th century. In the following years they dramatically unraveled the internal structure of the atom. As shown in the next chapter, the effort to understand the nature of cathode rays created a direct link to the discovery of that structure. Then, through a degree of serendipity, indirect observation of the atom's behavior, and much painstaking research, the "pieces" of the atom were discovered, leading scientists to figure out their peculiar, unique, and secret laws.

Chapter Five

Quantum Leap

"The atom revealed its secrets only gradually, like a cleverly written play, allowing sparks of knowledge briefly to illuminate corners of the story but jumbling scenes and inferences so that full realization came only at the end, after many members of the audience had given up hope of understanding."

ROBERT P. CREASE and CHARLES C. MANN
The Second Creation (1986)

From a distance, beach sand appears to be one piece of homogeneous material. Even though human eyesight is incapable of discerning each grain of sand from a distance away, no one would claim that the beach is a single entity or mass without constituent pieces. Likewise, despite our inability to see atoms, we have no basis for rejecting any particular configuration of them in the absence of proof. Seeing is not necessarily believing. If common sense suggests that the atom is made out of matter, then what kind of matter is it? Is that tiny piece of matter infinitely divisible, as Anaxagoras believed? Why? Into what? Is there a *smallest* piece of matter?

Röntgen's Accidental Discovery of X-Rays Begins the Search for Subatomic Particles

Rocks Beaming Out Rays . . . Plum Pudding

If a date could be set for the birth of nuclear physics, it would be November 8, 1895. On that day, Wilhelm Conrad Röntgen

(1845–1923), a physics professor at Wurzburg, Germany, was studying the emission of ultraviolet light from a cathode ray tube by directing the light at crystals of barium platinocyanide, which "fluoresces" under ultraviolet light. That is, the crystals seem to absorb the ultraviolet light and then return it by giving off their own light.

In the course of his experiments, Röntgen noticed that the rays emanating from the tube were penetrating *through* various substances and objects across the room, several feet away from the tube, and causing the crystals to fluoresce. After investigating this for the next several weeks and eliminating all other possible causes, he directed the rays at sensitive photographic plates and was able to register the image of his wife's bones in her hand, because the rays (which he called "X," the scientific symbol for the unknown) passed through the soft matter in her hand more readily than they passed through the bones. This was the first X-ray ever taken, and was to become one of the most famous images in science, not only because it led to the routine and widespread use of X-rays in medicine, dentistry and many other areas, but it was also a major step in determining the structure of the atom and the discovery of subatomic particles. In 1901 Röntgen became the first recipient of the Nobel Prize for Physics for his discovery of X-rays.

On January 7, 1896, two months after Röntgen discovered X-rays, the great French mathematician and scientist Henri Poincaré was sent photographs of these X-rays and was so amazed that he passed them on to two doctors and asked if they could duplicate Röntgen's work. On January 23, 1896, those Parisian physicians presented papers on the new phenomenon to the French Academy of Sciences. The physicist Antoine Henri Becquerel (1852–1908) was in the audience when the X-ray photos were shown. Becquerel, the third Becquerel in a row to occupy the physics chair at the Natural History Museum in Paris, was particularly interested in this subject because he'd been studying phosphorescence produced by a number of different types of rocks that glowed in the dark, and now wanted to test his hypothesis that there was some fundamental connection between X-rays and light. He experimented with different types of materials, and by February 24, 1896, he was confident enough about his conclusion that he presented a paper to the Academy in which he reported that phosphorescence from the rocks exposed to ultraviolet light from sunlight also emits the mysterious X-rays. However, when the sensitive photographic plates later recorded an image despite being stored in a dark drawer, he realized he was wrong. Becquerel promptly figured out it was the uranium in a piece of potassium uranium sulfate in the drawer that was spitting out some type of radiation all by itself—that is, without exposure to sunlight. Becquerel continued researching this phenomenon and published seven scientific papers on the subject in 1896.

As word spread around the world, physicists expanded their investigation of X-rays to the nature of the emanations from uranium. The source and nature of these emanations—labeled alpha particles or alpha rays—were completely unknown. The idea that a rock could beam out rays strong enough to penetrate photographic plates was contrary to everything scientists had believed, particularly the law of conservation of energy, which held that energy could never be created or destroyed. In contrast to the X-rays that Röntgen had created by electricity and the cathode ray tube, Becquerel's production of radiation seemed to come from *no place at all*. Thus, physicists pursued the search within the uranium atom to try to determine the source of the alpha particles.

In early 1897, Joseph John Thomson (1856–1940), director of the Cavendish Laboratory in Cambridge, England, the most prominent physics lab in Europe and perhaps the world at the time, was the first to discover that at least one type of the radiation being emitted from Röntgen's cathode ray tube consisted of a stream of small particles (which he called corpuscles, now known as electrons), each of which contained a negative charge. On April 29, 1897, Thomson presented a paper on this theory to the Royal Society, claiming that these particles were constituents of matter and were smaller than any known object, including the atom. This prompted an uproar, since the scientific community did not yet believe there could be anything smaller than an atom. Thomson speculated that Becquerel's alpha particles from the uranium were also electrons, but there was no evidence to support that theory. At the same time as Thomson was conducting his experiments, Heinrich Hertz (1857–1894), a professor at Bonn, Germany, reported that metal such as zinc that had been given a negative charge also gives up that charge when exposed to ultraviolet light. As a result of further work by Thomson, Hertz, and others, scientists became convinced of the existence of electrons.

The discovery of electrons then led to the proposition by some physicists that if there were negatively charged bits of matter in atoms, there must be positively charged matter to balance them out. But with little information on which to base a theory, their thinking was vague about such a positive counter force, and no one had yet suggested a clear or convincing image or description of the atom. Therefore, at the end of 1897 scientists had varying concepts of the atom: a "buzzing hive of thousands of electrons somehow held together by a positively charged glue" or "a spongy, doughy blob," or "a plum pudding," as J.J. Thomson suggested, with the electrons being the plums embedded in the nucleus.

Marie Curie Leads Way in Identifying Alpha Particles and Proving Existence of Radioactive Elements

Acid-Stained Hands and Leaky Roofs

In 1891, Marie Curie (1867–1934) (born Manya Sklodowska) had left her native Poland and arrived at Paris's *Gare du Nord* train station. The Eiffel Tower, the centerpiece of the 1889 World's Fair, loomed over the city as a symbol of modern technology along with electric lamps and internal combustion vehicles on the great boulevards. She'd come there to live near her sister Bronya—a doctor, who'd made it through school partly with Marie's financial support from her hard-earned and meager savings—reluctantly leaving their father and the rest of the family in Warsaw. Marie's hands soon became acid-stained for life as she performed chemistry experiments in Professor Gabriel Lippmann's lab at the Sorbonne in the hopes of a career in math and science.

In 1896 she received her certificate to teach secondary education to girls and in the following year decided to seek a Ph.D. in science, an achievement that no European woman had ever accomplished. Having been fascinated by the recent discoveries of X-rays by Röntgen and alpha particles by Becquerel, the thesis subject she chose was alpha particles. *What was the source of the energy that darkened Becquerel's photographic plates? What were the rays emanating from uranium?* Perhaps a stream of electrons, as Thomson had suggested, but of which nothing was known for certain. She selected this topic because the field was so new and could be pursued by doing experimental work in the laboratory, which she loved (as opposed to wading through books and articles in the library), and because it was a narrow and well-defined objective. As physicists groped for more detail and definitiveness in their concept and theory of the atom, Marie Curie began a friendly rivalry with Becquerel and a physicist named Ernest Rutherford to be the first to understand the source, nature, and behavior of the alpha particles that Becquerel had first discovered. This subject would dominate the rest of her life and forever tie her name to the history of nuclear physics.

Curie's work began in an unheated, damp, abandoned shed with a roof that leaked in the frequent Paris downpours. Her first notes were dated December 16, 1897, when she began using the piezo-quartz electrometer invented by her husband Pierre and his brother Jacques, a mineralogy professor. The piezoquartz electrometer measured the feeble electric current emanating from uranium and other substances. Pierre was head of the laboratory at the newly founded School of Industrial Physics and Chemistry in Paris, and worked closely with Marie on her research. However, it was Marie who first developed the theory that *the emission of the rays must be a phenomenon coming from within the atom of uranium itself.* That is, the rays were part of the matter, unlike X-rays, which are now known to be a form of electromagnetic radiation, and not something absorbed and returned in the form of phosphorescence. She coined the term "radioactivity" from the Latin "radius," meaning "ray." This simple hypothesis was to become Marie Curie's most important contribution to science, for it opened up the structure of the atom for others to build on, leading to the unfolding of that structure in the early 20th century.

Remarkably, on April 12, 1898, only a few months after she began this research, with no prior sources to rely on, Marie issued her first report, "Radiation Emitted by Compounds of Uranium and Thorium." The paper was presented to the French Academy of Sciences on her behalf by Professor Lippmann, since papers could only be presented by Academy members. On July 18, 1898, Marie and Pierre discovered the previously unknown element polonium (which she named after Poland), and on December 26 of that year, only a year after beginning to work on her thesis, the pair announced the existence of another new element, which they called radium. She determined that radium was a million times more radioactive than the uranium rock that had originally sparked all the research on radiation. Needless to say, Marie Curie earned her Ph.D. in science. In 1903 she and Pierre, together with Henri Becquerel, received the Nobel Prize for Physics for their discovery of radioactivity.

Rutherford Develops the Modern Picture of the Atom

Gold Leaf and Alpha Particles

New Zealander Ernest Rutherford (1871–1937) came to England in 1895 as the first foreign research student ever admitted to the Cavendish Laboratory. He left the Cavendish in 1898 to accept the physics chairmanship at McGill University in Montreal, where he continued his life's work studying alpha rays. During eighteen months of his nine years there, he worked with Frederick Soddy (1877–1956), a young chemist from Oxford, and discovered that elements that display radioactivity slowly change into other elements—further proof that the law of conservation of energy was not accurate and that chemical elements were not immutable. For example, the thorium with which they were working changed into helium. This work, published in 1902 in *Philosophical Magazine*, was the most significant contribution to the understanding of radioactivity since the Curies' discovery of radium four years earlier.

In 1903, as some scientists were still struggling with Thomson's "plum pudding" model of the atom, others suggested that atoms are made of varying numbers of one type of constituent called "dynamids." Then in 1904, the Japanese physicist Hantaro Nagaoka proposed that the electrons were orbiting in a "saturnian ring"—a positive sphere with a halo of mutually repellent electrons resembling the planet Saturn. While these theories of the structure of the atom were being considered, scientists determined in 1906 that the X-ray was a form of electromagnetic radiation of extremely short wavelength. Though Röntgen's discovery of X-rays led to the research on the atom and there would later turn out to be a direct relationship between that structure and X-rays, the questions now focused on *subatomic particles* such as electrons and alpha particles, not electromagnetic *radiation* such as X-rays, light, or radio waves.

In 1907 Rutherford returned to a post at the University of Manchester, England, and continued an ongoing debate he was having with Becquerel on the behavior of the alpha particles being spewed out of radioactive material. When Becquerel reported that his experiments indicated that alpha particles appeared to be bouncing off air molecules in their path, Rutherford began his own work to find the reason that occurs. With help from his assistants, including the German physicist Hans Geiger (1882–1945)—who later invented the Geiger counter, the first successful detector of alpha

Figure 5-1

RUTHERFORD'S APPARATUS FOR STUDYING ELEMENTS UNDER ALPHA-BOMBARDMENT

Thin layer of radioactive material releasing alpha particles

Adjustment screw

Microscope to observe flashes from protons impacting screen

Fluorescent screen

Released protons

Thin gold foil releasing protons as result of alpha-bombardment particles—Rutherford determined that one alpha particle in 8,000 rebounded when it slammed into a sheet of gold leaf.

Finally, in late November 1911 Ernest Rutherford determined that in order to deflect alpha particles, most of the atom would have to be concentrated into a tiny charged core in its center. He presented this theory to a session of the Manchester Literary and Philosophical Society on March 7, 1912. The deflection of alpha particles, proposed Rutherford, must be caused by something very small and very hard in the atoms of the material against which such particles collide, such as the sheet of gold with which Rutherford had been experimenting. He rejected Thomson's "plum pudding" model of the atom because electrons wouldn't bounce off it, and instead described the atom as *"a central electric charge concentrated at a point"* (referring to the nucleus) with *"a uniform spherical distribution of opposite electricity"* (referring to the electrons). It doesn't matter which charge is called positive and which negative—the point is that they're opposite charges. The "uniform spherical distribution" of electrons is traditionally considered negative.

Rutherford's public statement of his description marks the beginning of the modern understanding of the structure of the atom. He was first to the finish line, ahead of Curie and Becquerel. We now know that the alpha particles that Becquerel first discovered are the nuclei of ordinary helium atoms and are catapulted from decaying radioactive atoms at speeds of over 10,000 miles per second. For Rutherford's picture of the atom, he received many honors and was knighted in 1914, but he was never awarded the Nobel Prize for Physics.

Although Rutherford's work provided the basis for the modern concept of the atom, serious questions remained. A number of physicists, including Rutherford himself, soon realized that if he were correct, there would have to be a reason that the orbiting electrons would not be sucked into the positively charged nucleus. In other words, Rutherford's theory was groundbreaking, but his atom was both mechanically and electromagnetically unstable. Thus, there began a concerted effort to find some type of balancing mechanism within the atom.

Bohr Determines That Fixed Quantities of Energy Allow Electrons to Maintain Their Orbits

Honeymoon Missed . . . Collecting Tolls in Light Quanta

A native of Denmark, Niels Bohr (1885–1962) came to England on a scholarship in 1911 to work with J.J. Thomson at the Cavendish for one year. He'd studied Thomson's "plum pudding" nucleus and, like Rutherford, thought it could not possibly be correct. Shortly after arriving, Bohr had his first opportunity to engage Thomson in a discussion about his concept of the nucleus, but the Dane's poor grasp of the English language limited his ability to express himself, and he blurted out to the famous scientist, "This is incorrect!" Because of their differing concepts of the atom, Bohr's stay at the Cavendish was not pleasant, and he arranged to begin working for Rutherford at Manchester in March 1912, the same month that Rutherford presented his now-famous paper on the atom's structure.

His association with Rutherford proved fruitful, as the two shared a similar concept of the relationship between the electrons and the nucleus, including the fact that until someone adequately addressed the *instability* of the atom, no one would accept the picture that Rutherford had proposed. In the late spring of 1912, just a few months after associating with Rutherford, Bohr suggested that a theory about light developed by Max Planck and Albert Einstein might apply to the atom. Planck and Einstein had said that light consists of discrete amounts of energy called *quanta*. Bohr then speculated that quantization might be a fundamental property of *all* energy. If so, this could provide an explanation for the atom's stability. *Could fixed quantities of energy be related to fixed electron orbits around the nucleus?* Bohr set out to answer this question.

He returned to Copenhagen from Manchester in the summer of 1912 and married his fiancée, Margrethe Norlund. They canceled the honeymoon they'd planned so Bohr could spend the time developing his theory. Seven months later, in February 1913, Bohr was still struggling with how to describe this theory from both a conceptual and mathematical standpoint when a friend suggested that he look into a formula developed by a Swiss schoolteacher, Johann Balmer (1825–1898), which gives the frequency of light emitted by atoms. Bohr immediately realized that a combination of the Planck/Einstein quanta with this formula would provide the information he needed to describe the behavior of electrons orbiting the nucleus. His application of the formula to quanta demonstrated that:

- Electrons must exist in fixed arrangements, like satellites orbiting the globe at specified distances away from it.
- An electron leaping from one orbit to another results in either an increase in or the release of a "quantum" of energy.

An atom does not emit radiation while it is in one of its stable states, but only when it makes a transition from one state to another. That energy is in the form of electromagnetic radiation, some of which can be seen in the form of light. As explained by Robert Crease and Charles Mann in *The Second Creation*, "Quanta of light are the tolls collected or paid by electrons as they jump about their permitted places in the atom. The spectral lines described by Balmer's formula are a set of subatomic hops, skips, and jumps, the jitterbug moves of hydrogen's lonely electron." The atom neither absorbs nor emits radiation continuously, but only in "quantum leaps," a term commonly used outside of physics today that owes its origin to Niels Bohr's theory.

On March 6, 1913, almost one year to the day after Rutherford had described the atom as a central electric charge with a uniform spherical distribution of opposite electricity, his assistant and colleague, Niels Bohr, published his own theory in the *Philosophical Magazine* issues of July, September, and November 1913, and provided the solution to the major problem left open by Rutherford. The Rutherford-Bohr concept of the atom was born.

Bohr's work was confirmed later that year by the English physicist Henry Moseley (1887–1915) through a series of experiments that proved that the high-energy light given off when electrons move from a high to a low energy state was in the form of the X-rays originally discovered by Röntgen in 1895—thus,

Figure 5-2

RUTHERFORD-BOHR MODEL OF A CARBON ATOM

Nucleus

Electron

Orbital Path
of Electrons

the earlier-mentioned connection between X-rays and electrons—and that there's a mathematical correlation (known as Moseley's law) between the X-ray frequency of each element and its atomic number. If the frequency of the X-rays is known, the amount of the charge in the nucleus can be determined. This measure of X-ray energy, which is specific for each of the elements from which it emanates, is commonly used today to detect elemental "fingerprints" in any substance, such as a fluid or rock of unknown composition.

As mentioned in Chapter 4, the 19th-century chemists Avogadro and Cannizzaro figured out that atoms bind together into "firm unions" or molecules. However, it was not until the Rutherford-Bohr model of the atom was developed that we understood exactly how this chemical bonding occurs. The association of atoms into molecules occurs when there are "unpaired" electrons in the atom's outer shell or outer orbit. As a result of the tendency of electrons to distribute themselves among atoms so that the total energy of a group of atoms is lower than the sum of the component atoms, electrons are shared by atoms and thereby form molecules. The number of unpaired electrons in the outer shell is called the *valence* of the atom, and the bonding or sharing of electrons is called *covalency.*

Once again, if an atom were blown up to the size of the Superdome in New Orleans, the nucleus would be the size of a pearl in

Figure 5-3

MOLECULES

"Cloud" area where electrons are shared by atoms

Oxygen atom

Hydrogen atom

105°

Hydrogen atom

Polyethylene, a common plastic, consists of ethylene, the molecule shown above, which is made of two carbon atoms and four hydrogen atoms ($C_2 H_4$)

Water (H_2O) contains a central atom of oxygen flanked by two smaller atoms of hydrogen

the center of the structure, with the electrons orbiting at the outer perimeter of the building at an average speed of about two thousand miles per second, depending on the particular shell or orbit. Scaled down somewhat, if the nucleus were enlarged to the size of the period at the end of this sentence, the electrons revolving around it would be about seventy-five feet away, with that enormous amount of totally empty space between the dot and those electrons. The atom is virtually empty. Its size is only a description of the space or "cloud" through which its electrons vibrate and pass at that great speed. The diameter of the nucleus is about one hundred thousandth of the diameter of the whole atom. The volume of the nucleus is one trillionth of that of the whole atom. To put this in another perspective, if the Earth were enlarged to the incredible size of *93 million miles in diameter* (the distance from here to the sun), the nucleus of even the largest atom, increased proportionately, would still be microscopic. The basic characteristics of the atom seem so bizarre and foreign to everyday experience that it is difficult to conceptualize. This fundamental form of matter—fleeting, transient, ephemeral, surreal—is like an illusion. Yet the manifestation of the electrical charges is our tangible reality.

In 1921, the University of Copenhagen created the Institute of Theoretical Physics for Niels Bohr, and he served as its director for the remainder of his life. The institute soon became the world capital of atomic physics, attracting leading researchers and lecturers from the world over. In 1922 Bohr received the Nobel Prize for Physics for his work on atomic theory.

Following publication of Bohr's quantum theory, hundreds of theoretical and experimental physicists at the Cavendish and later at Bohr's institute delved deeper and deeper into the structure and workings of the atom—like archaeologists carefully chipping layers from a delicate fossil. Theories were constructed, then thrown into the worldwide academic arena to see if they could withstand laboratory experiments and critical analysis. Fierce intellectual battles ensued among the greatest names in physics, played out at laboratories, in correspondence, and in the pages of scientific journals. *Is the electron a wave or a particle? What causes electrons to change orbits? What forces keep the nucleus intact? Is there another particle in the nucleus?* During the 1920's a new physical interpretation of the atom was developed, called quantum mechanics, which linked such physical principles to Bohr's and others' mathematical descriptions.

One major aspect of the research performed in the succeeding years was directed at a large uncharged ("neutral") particle suspected to be present in the nucleus. Rutherford had predicted the existence of such "neutrons" as early as 1920, and conducted a series of experiments at Cambridge in an attempt to find them. However, these experiments were ultimately inconclusive, and his idea was pushed aside. But in 1932 Rutherford's former student James Chadwick (1891–1974), who was working at Cambridge University and well aware of Rutherford's prediction, confirmed the existence of the neutron in the nucleus one year after Irène Curie and Frédéric Joliot, the daughter and son-in-law of Marie Curie, handed Chadwick a decisive clue. They reported the discovery of a new type of high-energy but *uncharged* "radiation" made by bombarding beryllium with alpha particles. Curie and Joliot mistook this for gamma rays, thus leaving Chadwick to discover the elusive neutron, for which he won the Nobel Prize for Physics in 1935. Chadwick's discovery of the neutron not only completed the picture of the atom's basic structure consisting of three "pieces," it would also become an essential step in harnessing and releasing the enormous power hidden in the nucleus.

On July 4, 1934, Marie Curie died at the age of 67, a victim of nearly four decades of overexposure to radioactive elements. By this time, her legacy was evident all around her, as the discoveries she'd made rippled forward with increasing momentum into the 1930's, and as knowledge of the atom unfolded. The discovery of the neutron, together with several other discoveries that took place in the 1930's, dramatically and permanently transformed this pure laboratory science into a field with practical application. The world was rapidly approaching the nuclear age.

Bohr's institute remained a focal point for theoretical physics through the 1930's, and played a pivotal role in understanding how to split the atom, as we'll see in Chapter 6. When World War II erupted in 1939, Denmark was overrun and occupied by the Germans, and in 1943, risking his life because of his Jewish ancestry, Bohr escaped to England, then worked for two years on the Manhattan Project in Los Alamos, New Mexico, as part of a British research group.

The work conducted by physicists since the time that Bohr accurately described the atom in 1913 fills a hundred biographies and scores of texts with material well beyond the scope of this book. Researchers have not only "created" new chemical elements, they've broken the three constituent subatomic parts of the atom into even smaller particles. As for new elements, we're up to number 112 (unnamed as of 1997), consisting of 112 protons and 165 neutrons, thus, 20 elements beyond uranium, the heaviest *naturally occurring* element. Element 112 was formed in the lab by slamming zinc ions into a lead nucleus. However, like most of the other artificially created heavy substances, this new element is highly unstable and decays in milliseconds. With the exception of a few heavy elements that occur naturally in trace amounts, elements with atomic numbers greater than 92 (uranium) were discovered as synthetic products of particle accelerators at nuclear research laboratories or as debris from nuclear explosions. They are highly radioactive and have been synthesized in tiny quantities, usually only a few atoms at a time.

Regarding the further breakdown of the electron, proton, and neutron, researchers have discovered a proliferation of *sub-subatomic* particles. There are actually *hundreds of particles*, but they fall into two basic constituents of matter: *six quarks and six leptons*. For example, the lepton category includes the electron, muon, tau, and their associated neutrinos. The proton is composed of three kinds of quarks, which are generally considered indivisible. The majority of these hundreds of particles decay into other particles within tiny fractions of a second. This fundamental structure of matter into six quarks and six leptons is called the Standard Model. However, in January 1996, particle physicists at the world's most powerful accelerator, the Tevatron at the Fermi National Accelerator Laboratory in Batavia, Illinois, found significant evidence that quarks might be further divisible—that is, they might have some type of internal structure.

Was Anaxagoras right? Is matter infinitely divisible? Although that question can't yet be answered with certainty, it *is* known with certainty that the subatomic particles discovered so far have specific attributes: electric charge, mass, specified quanta of energy, angular momentum, spin (direction of rotation), and velocity of such rotation.

Ernest Rutherford predicted that "the forces holding the nucleus together will not be found in this generation or the next . . . or for many hundreds of years, because the constitution of the atom is, of course, the great problem that lies at the base of all physics and chemistry . . ." Though we will not delve further into the story of the creation of new atoms or the research into quarks, leptons, or angular momentum, we will explore how physicists came to understand "the forces holding the nucleus together." In the course of their research into the atom, they discovered something so incredible and with such far-reaching ramifications for all of us that it must be part of every discussion of the structure of the atom: the awesome power concealed and contained within the atomic nucleus. The bomb and nuclear power, with their mixed legacy of fear and wonder. Destruction and promise.

Therefore, the next chapter looks at how scientists identified and harnessed the unbelievably immense force and energy in this surreal, illusory, impossible, invisible, textureless, electrically charged "pearl" suspended in the central point of the Superdome.

Chapter Six

The Crack of Doom

"I firmly believe that before many centuries more, science will be the master of man. The engines he will have invented will be beyond his strength to control. Some day science shall have the existence of mankind in its power, and the human race commit suicide by blowing up the world."

HENRY ADAMS (1862)

"The atom stands as a monument to the wisdom of the Human Race. One day it may stand as a tombstone to its folly."

J.G. FEINBERG
The Story of Atomic Theory and Atomic Energy (1960)

"The world is very different now. For man holds in his mortal hands the power to abolish all forms of human poverty and all forms of human life."

JOHN F. KENNEDY
Inaugural Address (1961)

The free world rejoiced when the Berlin Wall came tumbling down in 1989, and we heaved a collective sigh of relief on September 12, 1990, when the Cold War came to a formal end with the "Treaty on the Final Settlement with Respect to Germany." Signed by the United States, France, England, and the Soviet Union, this agreement allowed German reunification. We further relaxed in July 1991 when the START Treaty (Strategic Arms Reduction Talks) was signed and implemented, signaling the beginning of the dismantling of tens of thousands of nuclear warheads. These developments reduced the immediate threat of possible nuclear holocaust. However, as mere political changes are of questionable strength and duration, agreements on paper can never fully eliminate the uncertainty and anxiety concerning nuclear weapons and their potential to destroy civilization. Truces among nations that have conflicting political systems and otherwise compete in this world are not institutions. They are, at best, a temporary and highly fragile state of affairs.

Thus, the threat of nuclear winter still hovers like a dark cloud. Ever since the secret of the mushroom cloud escaped from the atom's nucleus, the destinies of the atomic bomb and the human race became intertwined forever. The specter of that cloud will be with us as long as there is political uncertainty in the world. As long as there are nations, history has proven that civilizations will arise and vanish, and superpowers will be born and die in the centuries and millennia ahead. In a hundred years, who will threaten to use the atomic bomb? As Theodore Zeldin wrote in *An Intimate History of Humanity:*

> If a film were to be made compressing into a couple of hours all that . . . happened . . . the world would look like the moon, grey and desolate, remarkable only because there are a few craters on it. The craters are civilizations—thirty-four major ones so far—each exploding then dying out, having briefly lit up parts of the globe, but never all of it; some last for a few hundred years, others for a couple of thousand . . . All civilizations, so far, have decayed and died, however magnificent they have been in their glory, however difficult it is to believe that they can vanish . . .

We came to discover and unleash the secret symbolized by the mushroom cloud when our quest to fully understand the structure and function of the atom intersected with the ever-changing political and military history of the world.

Scientists Tie the Sun's Energy to the Atom's Nucleus

Missing Mass and the Key to All Life

Nineteenth-century scientists accepted the "contractive hypothesis" of the Scottish physicist William Thomson (1824–1907) (known as Lord Kelvin), which held that the Sun obtains its energy supply from a slow but steady shrinking of its giant body. However, this theory only accounted for enough energy to support the Sun's radiation for about 20 million years. Thus, when biologists, geologists, and paleontologists discovered that evolution of life on Earth had begun well before 20 million years ago, Kelvin's theory was abandoned.

Henri Becquerel's 1896 discovery of spontaneous energy emanating from radioactive materials provided the first clue to the Sun's immense energy, but it was not until 10 years later, when Einstein introduced his famous formula ($E=mc^2$), that the true relationship between matter and energy and the true source of the Sun's light and heat was understood. *Energy equals mass times the speed of light squared.* With this equation, Einstein was proposing that matter ("mass") and energy are both *destructible.* In fact, one can be *converted* into the other. This was directly contrary to the supposedly inviolate Law of the Conservation of Energy—formulated in 1847 by the German physicist, Hermann Ludwig Ferdinand von Helmholtz—that stated that neither matter nor energy could be destroyed or created. Einstein's theory raised the possibility that the Sun's matter was being destroyed in some way and was being converted into energy in the form of heat and light.

Mass is the *quantity of matter* in the particular object, as mentioned in the discussion of Newton's laws of physics. That is, it reflects the actual number of protons and neutrons in the atoms that make up the object, as we saw in Chapter 5. Unlike the *weight* of the object, mass is not subject to change and doesn't depend on the gravitational pull of the Earth or some other body. Therefore, physics uses this universal measure. On a larger physical scale, mass is determined by the amount of force required to produce a given acceleration, as suggested by Newton's Second Law of Motion (acceleration is directly proportional to the force).

According to Einstein's equation, the amount of energy stored in the mass of any matter is incredibly huge. As stated, the energy (E) equals the particular unit of mass (m) multiplied *by the speed of light* (c) *squared* (which is 186,000 miles per second). In contrast to atomic energy, if you were to burn 1/15th of an ounce of hydrogen gas, it would produce enough energy for a 100-watt light bulb to burn for about 40 minutes. But if you convert the mass (i.e., all the atoms) of the same amount of hydrogen gas into energy, it would result in enough energy to power the same light bulb for *56,000 years*. Similarly, one pound of the radioactive element uranium yields as much energy as three million pounds of coal. Burning is simply a chemical reaction in which energy that is stored in the molecular structure is released. In other words, burning is a re-alignment of the *chemical* bonds between electrons, not a release of energy stored in the atom's nucleus.

Scientists came to realize that the Sun was creating its energy by *fusing* hydrogen atoms together to form helium atoms—four hydrogen atoms fuse together to form one helium atom. When this happens—as is occurring continuously in every star throughout the universe—mass is converted into energy. Put simply, when four hydrogen atoms combine to make one helium atom, that resulting helium atom has slightly *less* than four times the mass of one hydrogen atom. The "missing mass" has been converted to energy, which we see (and depend on to exist) in the form of light and heat. This combination, or fusion, of hydrogen atoms into helium atoms is due to the incredible heat of the Sun's core—18,000,000 to 36,000,000°F. This phenomenon—the conversion of mass to energy—is of the ultimate significance. It is the single factor that accounts for all life on this planet.

Ironically, Newton actually foreshadowed Einstein's equation in his famous book *Opticks* when he wrote, "Are not gross Bodies and Light convertible into one another, and may not Bodies receive much of their Activity from the particles of Light which enter their Composition?" However, the idea that this energy could somehow be captured and used, as suggested by Einstein's 1905 proposal, remained a curious and intriguing speculation for 27 years because there was no way to test it. That is, there was no way to duplicate those enormous temperatures. In fact, soon after Einstein discovered that E=mc², he noted, "There is not the slightest indication that energy will ever be obtainable."

Scientists Determine That the Energy of the Atom's Nucleus Is Obtainable

Understanding the Nuclear Fluid

In 1928 the Russian-born nuclear physicist George Gamow (1904–1968) developed the first quantum theory of radioactivity, which earned him a fellowship at Niels Bohr's Copenhagen Institute of Theoretical Physics. While there (1928–1930) he formulated the "liquid droplet" model of an atomic nucleus according to which different nuclei are considered the minute droplets of a universal "nuclear fluid." He based his theory on the assumption that the forces acting between the constituent parts of an atomic nucleus are similar to those acting between the molecules of an ordinary liquid. Gamow noted that the sphere is the geometrical figure that possesses the smallest surface for a given total volume. If there is no external force acting on a drop of liquid (such as gravity pulling it down onto a flat surface), the natural surface tension or surface energy will work in drawing and keeping the constituent parts together in the form of a sphere. Likewise, he surmised that it would take "extra" energy to overcome the force holding the nucleus of an atom together, called the "strong nuclear force." He further pointed out that there is an energy opposing that "surface tension" effect, namely, the repulsion among and between the positively charged protons in the nucleus. The nucleus is a cohesive and tightly closed body of particles despite the natural tendency of the protons to split apart like the positive ends of two magnets. This indicates there is a natural equilibrium in every atomic nucleus. In other words, since the nuclei do not spontaneously split into smaller pieces or fuse together with other nuclei, Gamow reasoned that the strong nuclear force holding together the "nuclear droplet" *exceeds* the natural internal repulsion force of the nucleus.

Figure 6-1

Nuclear "Droplet" With Forces

Strong nuclear force
keeps nucleus
together

But natural
repulsive force
exists among
positively charged
protons

The electric repulsion force among the protons and the strong nuclear force holding the nucleus intact each have a specific mathematical value. By applying those values to the nucleus, Gamow demonstrated *theoretically* that splitting the "nuclear droplet" into two halves would result in an *increase* in the total surface energy, while the electric repulsion force among the protons would be *reduced*. Thus, nuclear physicists began to better understand what happens to that reduction or release of energy from the nucleus when fusion takes place in the Sun and began to seriously consider that such energy could be captured through fusion or fission (that is, splitting the atom's nucleus) and put to some practical use.

As discussed in Chapters 4 and 5, there are 92 naturally occurring elements of increasingly larger atomic weight and mass. The nuclei of hydrogen, helium, and some other light elements were formed relatively soon after the Big Bang (see Part Four), when the universe was extremely hot and dense, while other nuclei and atoms were formed as a result of stars' fusion reactions and as a result of stars exploding at the end of their normal life cycles. In the present "cool" universe, all these elements are in a very stable state. In Chapter 5 we saw how electrons transfer freely to the outer orbits of adjacent atoms when they form molecules. In contrast to this phenomenon, the strong nuclear force holding the nucleus together is much stronger and more complicated. The strong nuclear force is one of the four basic forces of nature, along with gravity, the electromagnetic force, and the weak nuclear force, which connects the orbiting electrons to the nucleus and is the force responsible for holding together atoms that combine to form molecules.

Gravity and electromagnetism act over great distances across the universe, while the strong nuclear force can only be detected in a laboratory since it operates only on a tiny scale in the atom's nucleus, binding protons and neutrons together.

Despite the extremely localized effect of the strong nuclear force, it would take a large amount of energy to destabilize the nucleus and create fusion or fission—much more "activation" energy than is needed to heat an ignitable substance before it will burn or to charge a high explosive in order to detonate it. For fusion to occur, nuclei must be propelled toward each other to overcome their inherently repulsive forces. For fission to take place, a violent collision is usually needed to break up a nucleus, with the exception of naturally radioactive atoms that undergo spontaneous fission. Such radioactive atoms contribute substantially to the internal heat of the Earth and decay at rates so statistically reliable that they are used as natural clocks and have invaluable application in geology to determine the age of rocks and other substances on Earth.

By the early 1930's—based largely on Gamow's nuclear droplet model—scientists realized that although it would be impossible to duplicate the Sun's *fusion* process caused by its enormous interior temperature, they might be able to harness the huge amount of energy predicted by Einstein's formula through *fission*. Harnessing any appreciable amount of energy through fission is more complicated than merely knocking off part of the nucleus. Henri Becquerel and Ernest Rutherford had been chipping protons off the nuclei of atoms since early in the century by bombarding them with alpha particles from a radioactive source, and their labs didn't blow up. Similarly, in 1919, Rutherford had succeeded in splitting nitrogen atoms into hydrogen and oxygen atoms by bombarding them with alpha particles. As we entered the 1930's, physicists pursued the idea of somehow striking enough atomic nuclei with such force that fission would take place on a scale sufficient to release and hopefully capture the energy escaping from the "nuclear droplet."

The Nucleus Is Split and the World Enters the Atomic Age

Escape From Italy

With the deeper understanding of the forces holding the nucleus together, scientists realized they would need a device that would artificially project protons at nuclei with great speed. They knew it would be very difficult for such protons to penetrate the natural barrier of electric repulsion resulting from the protons in the nucleus. Such a successful violent collision first occurred in 1932 when the British physicist Sir John D. Cockcroft (1897–1967) and the Irish physicist Ernest Walton (1903–1995), working with Ernest Rutherford at Cambridge's Cavendish Laboratory, created the first atom-smasher, which became known as the Cockcroft-Walton generator. They accomplished the first complete artificial splitting of an atomic nucleus (lithium nuclei) by bombarding it with protons and causing fission of lithium into two helium atoms. The energy gain was small, but it was the first demonstration of Einstein's prediction that energy would result from the conversion or loss of mass ($E=mc^2$), and might be captured and used.

Cockcroft and Walton were later awarded the Nobel Prize for Physics for this work. However, the Cockcroft-Walton generator and similar equipment that succeeded it were limited by the haphazard and unpredictable manner in which fission was accomplished. Without repetitive releases of energy from splitting nuclei, more energy is expended in accelerating the projectiles used to split the nucleus than is obtained from the fission itself.

As mentioned in Chapter 5, 1932 was also the year that James Chadwick discovered the existence of the neutron in the atom's nucleus—a particle with mass similar to the proton, but with no electrical charge. Scientists quickly realized that the ideal "atomic bullet" to split the nucleus would be the neutron, not the proton, since the neutron would not be electrically repelled by the positively charged nucleus. Thus, another turning point in making atomic energy a reality occurred in 1935 when the Italian physicist Enrico Fermi (1901–1954) conceived the idea that: (a) neutrons could be obtained from radioactive substances; (b) compounds rich in hydrogen (such as water) slowed down the neutrons being obtained from such radioactive compounds; and (c) when such neutrons were slowed, the likelihood that they'd be *captured* by an atomic nucleus *target* was increased. Fermi reasoned that when the large mass of the neutron impacts and binds with the target nucleus, the natural repulsive force between the protons would become greater than the strong nuclear force keeping together the nuclear droplet, thereby causing the nucleus to split apart. The incoming neutron could be sufficiently disruptive even to split a nucleus with many protons and large potential energy (say, the nucleus of a uranium atom) into two particles of approximately equal size, and release that energy. In 1938 Fermi was awarded the Nobel Prize for Physics for his work in neutron bombardment with slow neutrons. When Mussolini's fascist government granted Fermi permission to go to Sweden to receive the award, Fermi and his family carried out their plan to permanently leave Italy. They soon settled in the United States, where Fermi became a professor at Columbia University.

The Community of Nuclear Physicists Is Forced to Shift Its Focus

Yours Very Truly . . . Albert Einstein

As 1939 began, fission of a large nucleus into two halves and the release and harnessing of its enormous energy remained a theory. If it could be achieved, nuclear physicists envisioned using atomic energy for peaceful, industrial purposes for the good of mankind. Yet scientists realized that they must achieve a continuous and prolonged series of neutron bombardments in which the neutrons of such large nuclei actually become the projectiles to disrupt other nearby large nuclei. Without such a "chain reaction," more energy would be used in accelerating the projectile than would be obtained from the fission, as Cockcroft and Walton had experienced. This was the year World War II broke out in Europe, and it is far from a coincidence that the creation of an atomic bomb suddenly displaced the vision of harnessing atomic energy for peaceful use.

The German mathematical physicist Lise Meitner (1878–1968) figured out how to achieve the fission and chain reaction of a heavy element, thereby becoming a key figure in the development of the atomic bomb. After conducting experiments in Germany similar to Fermi's neutron bombardment, Meitner and two of her colleagues realized they had actually split uranium atoms into several parts and created a chain reaction on a small scale. In January 1939 she secretly left Germany and went to Stockholm, where she and her nephew Otto R. Frisch (1904–1979), a professor at the Institute of Theoretical Physics of the University of Copenhagen, working under Niels Bohr, then submitted a letter to the editor (titled "Disintegration of Uranium by Neutrons: New Types of Nuclear Reaction") in the most prestigious British science journal, *Nature*, explaining Meitner's experiment. Their report appeared in the February 11, 1939, issue, and was the first work to use the term "fission" to describe the division of uranium into lighter elements:

Figure 6-2

FISSION AND CHAIN REACTION

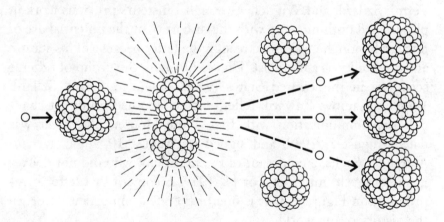

One neutron impacts uranium nucleus and produces two fragments of the original atom → and also produces an average of three neutrons → which then induce fission in other nuclei–resulting in instant release of energy in the form of radiation

Upon splitting a uranium nucleus (which contains 54 more neutrons than protons), two new nuclei (such as barium atoms with 26 extra neutrons) would be formed, thereby resulting in a discharge of the neutron oversupply, thus further resulting in a chain reaction when those neutrons smash into the nuclei of other uranium atoms and continue the fission of the nuclei of such atoms.

Realizing the significance of Meitner's work, Fermi and Bohr engaged in studies and experiments in 1939 in the United States that further indicated the feasibility of a nuclear chain reaction. Later that year, Fermi and two other physicists at Columbia University, the Hungarians Leo Szilard (1898–1964) and Eugene Wigner (1902–1995) brought the work on fission to Albert Einstein's attention. The U.S. government feared being drawn directly into the war and continued to support its allies against Hitler. Convinced the Nazis would be in possession of atomic power in the near future and would use it to subdue the rest of the world, Einstein agreed to sign a letter to President Roosevelt that was drafted by Fermi, Szilard, and Wigner. Although Einstein's previous work in physics had nothing to do with the discovery or the potential use of nuclear fission, he was asked to sign the letter because of his stature and credibility as a physicist. Unlike Einstein, the group of refugee scientists in the United States (which included Fermi, Szilard, Wigner, Gamow, Edward Teller, and others) had no direct channels of communication with the White House. The letter was dated August 2, 1939, and signed on August 10, then given to Alexander Sachs, a confidant of the president, who did not deliver it to Roosevelt until October 11, 1939. A copy of that letter is reproduced in Figure 6-3. It is one of the most important letters in the history of the world.

As a result of this letter, Roosevelt formed a committee charged with examining the prospects of the military use of atomic energy. After the committee reported to the President that it was possible to use such energy in atomic bombs, the U.S. Army and Navy made their first grant for atomic research in February 1940 in the amount of $6,000. Though a meager sum even by 1940 standards, this marked the first involvement of the U.S. military in atomic energy

Figure 6-3

EINSTEIN'S LETTER TO FDR

Albert Einstein
Old Grove Rd.
Nassau Point
Peconic, Long Island

August 2nd, 1939

F.D. Roosevelt,
President of the United States,
White House
Washington, D.C.

Sir:

Some recent work by E. Fermi and L. Szilard, which has been communicated to me in manuscript, leads me to expect that the element uranium may be turned into a new and important source of energy in the immediate future. Certain aspects of the situation which has arisen seem to call for watchfulness and, if necessary, quick action on the part of the Administration. I believe therefore that it is my duty to bring to your attention the following facts and recommendations:

In the course of the last four months it has been made probable - through the work of Joliot in France as well as Fermi and Szilard in America - that it may become possible to set up a nuclear chain reaction in a large mass of uranium, by which vast amounts of power and large quantities of new radium-like elements would be generated. Now it appears almost certain that this could be achieved in the immediate future.

This new phenomenon would also lead to the construction of bombs, and it is conceivable - though much less certain - that extremely powerful bombs of a new type may thus be constructed. A single bomb of this type, carried by boat and exploded in a port, might very well destroy the whole port together with some of the surrounding territory. However, such bombs might very well prove to be too heavy for transportation by air.

The United States has only very poor ores of uranium in moderate quantities. There is some good ore in Canada and the former Czechoslovakia, while the most important source of uranium is Belgian Congo.

In view of this situation you may think it desirable to have some permanent contact maintained between the Administration and the group of physicists working on chain reactions in America. One possible way of achieving this might be for you to entrust with this task a person who has your confidence and who could perhaps serve in an inofficial capacity. His task might comprise the following:

a) to approach Government Departments, keep them informed of the further development, and put forward recommendations for Government action, giving particular attention to the problem of securing a supply of uranium ore for the Untied States;

b) to speed up the experimental work, which is at present being carried on within the limits of the budgets of University laboratories, by providing funds, if such funds be required, through his contacts with private persons who are willing to make contributions for this cause, and perhaps also by obtaining the co-operation of industrial laboratories which have the necessary equipment.

I understand that Germany has actually stopped the sale of uranium from the Czechoslovakian mines which she has taken over. That she should have taken such early action might perhaps be understood on the ground that the son of the German Under-Secretary of State, von Weizsücker, is attached to the Kaisor-Wilhom-Inotitut in Berlin where some of the American work on uranium is now being repeated.

Yours very truly,

A. Einstein

(Albert Einstein)

research. After signing the letter, Einstein had no further connection with research or development of atomic energy or the atomic bomb.

The Manhattan Project Is Initiated

Little Boy and Fat Man . . . 210,000 Dead

In August 1942, Roosevelt charged U.S. Army and Brigadier General Leslie R. Groves with the responsibility of organizing the effort—with support from Great Britain and Canada—to harness nuclear energy for the military. This became known as the Manhattan Project because much of the research was being done by Enrico Fermi and others at Columbia University in Manhattan and because the Army's Manhattan District office initially organized the project. In October 1942, J. Robert Oppenheimer (1904–1967) was named director of the newly formed Atomic Research Laboratory. The son of a German immigrant, Oppenheimer had graduated from Harvard in 1925, then did research in physics at the Cavendish Laboratory at the University of Cambridge under Ernest Rutherford during the period of pioneering studies on atomic structure. He received his Ph.D. in 1927, then began teaching physics at Berkeley. In 1939, when Germany invaded Poland and Einstein signed the letter to President Roosevelt, Oppenheimer began working with the small circle of nuclear physicists who were aware of the potential of devising a nuclear bomb. His special area of expertise—separating uranium-235 from natural uranium, which is predominantly uranium-238—became a critical area of research. Under the cloak of secrecy, Oppenheimer assembled the best nuclear physicists available to work in various laboratories around the country, and at a facility in the New Mexico desert, called Los Alamos, to understand the implications of a chain reaction of heavy elements such as uranium to create fission. Oppenheimer's group included Enrico Fermi and these other individuals:

- Hans Albrecht Bethe (1906–)—German-born physicist who worked with Fermi in Italy in 1931, emigrated to the U.S. in

1934, and headed the Theoretical Physics Division at Los Alamos. In 1939 Bethe had calculated the specific amount of energy created and the specific nuclear process involved in the Sun's fusion.

- Edward Teller (1908–)—Hungarian-born American nuclear physicist who studied at Niels Bohr's Institute in Copenhagen, collaborated with Gamow on understanding the nuclear decay process, worked on Fermi's team at the University of Chicago, and was one of the first recruited by Oppenheimer to go to Los Alamos.
- Willard F. Libby (1908–1980)—American chemist who helped develop the method used to separate uranium-235 from uranium-238.
- William G. Penney (1909–1991)—British nuclear physicist who led the team of British scientists working on the Manhattan Project.
- Leo Szilard (1898–1964)—Hungarian-born American physicist who was instrumental in persuading Einstein to sign the letter to FDR, developed the first method to separate U-235 from U-238, worked closely with Fermi on the 1942 chain reaction at the University of Chicago, and continued to work on the Manhattan Project until the end of the war.
- Harold C. Urey (1893–1981)—American chemist who discovered deuterium (heavy hydrogen) in 1931, which advanced the development of fission, and who later directed the research at Columbia University that provided fundamental information on separating U-235 from U-238.

Assigned the task of producing a controlled, self-sustaining nuclear chain reaction, Fermi and his colleagues pursued this work on an abandoned squash court beneath the stands of Stagg Field at the University of Chicago. His group had calculated that approximately 54 tons of uranium interspersed with about 400 tons of graphite would be the optimum volume and combination of material to create such a chain reaction. On December 2, 1942, they were cautiously adding to the twelve-foot-high "pile" of uranium and graphite bricks when it came time for another test. As they slowly withdrew the cadmium control rods one by one, the meter that recorded neutron release and bombardment began to jump wildly. It reflected their success in obtaining a chain reaction—that is, the splitting of millions of uranium atoms. They reinserted the control rods to snuff out the activity. One of Fermi's assistants ran from the test to the telephone to inform Oppenheimer of the results. "The Italian navigator has just landed in the New World," he reported. This now-famous experiment represented the culmination of almost five decades of intense and continuous effort to fully understand the atom. Fermi and his group were the first people to demonstrate that Einstein's formula could have practical application, whether it be to provide needed energy in peacetime, or as a military weapon, or to "commit suicide by blowing up the world."

Once the chain reaction was moved from the realm of conjecture to the realm of the probable, secret plants were set up on a 70-square-mile area at Oak Ridge, Tennessee, and on an isolated 1,000-square-mile tract (named the Hanford Engineer Works) north of Pasco, Washington, on the banks of the Columbia River for access to its water to cool the enormous heat of the atomic piles to be constructed there. At Oak Ridge—where one of the buildings covered 44 acres—the uranium-235 atom needed for fission could be stripped from its natural companion, the predominant uranium-238 material. At the equally massive Hanford complex, the experimental manufacture of plutonium-239 was to take place. U-235 and plutonium-239 were earmarked as having the best potential for creating the explosive chain reaction necessary for a nuclear bomb. The population at Oak Ridge eventually reached 75,000, and at Hanford 50,000.

The product resulting from the work by the people at Oak Ridge, Hanford, and other locations was intended to be used at a secret facility 35 miles northwest of Santa Fe, New Mexico, on the flat and desolate expanse of the southwestern desert called Los Alamos. The site had been chosen by Oppenheimer for its very remoteness in November 1942, a month after he'd been selected to head the Manhattan Project. At Los Alamos, he had combined the efforts of the great physicists he'd brought together, including Fermi and the others listed above. The scientists at Los Alamos were given an awesome assignment: *to design the actual bomb to be powered by the fissionable material being manufactured at Oak Ridge and Hanford.*

One essential question they immediately faced was determining the amount of U-235 or plutonium in the bomb. That is, what is the "supercritical mass" of fissionable material? The time elapsing between the ejection of a neutron from a uranium atom that is being split and the resulting splitting of a second uranium atom is *one hundred millionth of a second.* If too little uranium was used, the neutrons would escape from such material without creating the necessary chain reaction. But if too much uranium was gathered together closely, it would instantaneously explode and destroy Los Alamos and all 5,000 people there. If the Los Alamos team could solve this and the other enormous problems they faced in designing the bomb, its force—with each *pound* of uranium or plutonium exploding with the power of 8,000 tons of dynamite—would stagger the imagination . . . and perhaps end the war.

After about two years of work at Oak Ridge and Hanford, enough fissionable material was produced to enable the Los Alamos group to begin to physically assemble the first atomic bomb—a device about six feet long, two feet in diameter and weighing four tons, and containing what they had calculated to be the supercritical mass of plutonium-239. Thus, about five and a half years after Roosevelt read Einstein's letter, and after the expenditure of two billion dollars and millions of man-hours, the countdown to test the physicists' invention began in the pre-dawn hours of July 16, 1945, at the remote Alamogordo Air Base in New Mexico. The bomb was placed on top of a 100-foot steel tower surrounded by scientific monitoring equipment. Robert Oppenheimer and his group were in the control room six miles away from the bomb site, while other scientists and observers huddled in bunkers and behind shelters and other fortifications ten miles from the site. Time minus 50 minutes, time minus 30, 20, 10, and so on until the voice over the speaker shouted, *"Now!"* Suddenly, a ball of fire shot skyward and the observers were flooded by a burst of golden, purple, gray, and blue light *one hundred times the intensity of the afternoon sun on a clear day.* The detonation was the equivalent of 20,000 tons of dynamite. Thirty seconds into the furious blast, several observers standing behind a shelter were knocked down by the rush of air, even though they were ten miles away from the explosion. General Thomas Farrell, the chief military officer at Los Alamos, later wrote in his report to the War Department that the explosion was "unprecedented, magnificent, beautiful, stupendous and terrifying." The steel tower was completely vaporized, and the surrounding desert sand for a radius of 800 yards was fused to glass.

As the mushroom cloud rose to 40,000 feet in the air, the people at Alamogordo that fateful day witnessed the shape that would influence much of history for decades and perhaps centuries. They wrote the last chapter of the story that joined the past to the future, for this was the culmination of humankind's fascinating 2,400-year-long struggle to gain insight into the ultimate piece of matter—Leucippus's particle, the very beginning of all matter. This tale ended with an entirely unexpected and extremely frightening twist. Through an amazing application of human reasoning power, a line of outstanding scientists dating back to Robert Boyle and the other 17th-century and 18th-century chemists not only figured out the potential within the atom, they also made Einstein's formula a prophecy as the energy within the atom was harnessed and its fantastic secret released, to be with us now, in peace and in war, forever.

On August 6, 1945, three weeks after the test at Los Alamos, a single B-29 bomber, the *Enola Gay*, carried a bomb containing U-235 across the Pacific Ocean and released it over Hiroshima, Japan, at 8:15 in the morning. Nicknamed "Little Boy," the weapon detonated 1,900 feet above the ground, destroying two-thirds of the city and killing 140,000 of its 350,000 inhabitants. A second weapon, nicknamed "Fat Man," similar to the plutonium bomb tested at Los Alamos, was scheduled to be dropped on Kokura, Japan, on August 11, but was moved up two days to avoid bad weather predicted for the area. The B-29 *Bock's Car*, which carried this bomb, spent ten minutes over Kokura without locating its target through the clouds, then proceeded to its secondary target, the city of Nagasaki. At 11:02 a.m., Fat Man destroyed half the city and killed 70,000 of its 270,000 people.

Part Three

The Principle of Relativity

In Focus

What could be more certain and absolute than the passage of time? At the age of 16, Albert Einstein posed a question to himself that led, ten years later, to his formulation of the special theory of relativity, which shows that the rate at which time passes is *not* certain or absolute. The concept of universal time must be replaced with a multitude of "personal" times. Time and space are no longer separate entities. They must be replaced with a fourth dimension: *space-time.*

Einstein's theory did not replace Newtonian physics. Special relativity only applies to an object moving at extreme speed in relation to another. Newton simply hadn't contemplated those conditions and didn't have the information about electromagnetic waves and the speed of light available to him. Ten years after Einstein discovered the special theory, he completed his development of the general theory of relativity—a new view of the source and effects of gravity throughout the universe. His impact in numerous other fields of science was also significant.

Einstein was one of the greatest intellects of all time. But he also immersed himself in the pressing social issues of his day, compassionately speaking out at every opportunity and making great personal sacrifices as he tried to make the world a more civil place. We will first look at the person behind the theory; then we will examine the theory that shook the very foundation of time.

Chapter Seven

Philosopher-Scientist

"How strange is the lot of us mortals! Each of us is here for a sojourn; for what purpose he knows not . . . The ideals which have lighted my way, and time after time have given me new courage to face life cheerfully, have been Kindness, Beauty and Truth."

ALBERT EINSTEIN
The World As I See It (1931)

Albert Einstein was a philosopher. He also happened to be an extraordinary physicist. We know about him today because of his theories about space, time, and gravitation, but he was equally passionate about freedom, the value and purpose of human life, good and evil, education, religion, politics, government, and pacifism. Einstein wrote or spoke publicly on such issues over two hundred times between 1915 and his death in 1955, despite the fact he shied away from his worldwide fame:

> Let every man be respected as an individual and no man idolized. It is an irony of fate that I myself have been the recipient of excessive admiration and reverence from my fellow-beings, through . . . no merit of my own. The cause of this may well be the desire . . . to understand the few ideas to which I have with my feeble powers attained . . .

Although Einstein devoted most of his life to science, he had a profound interest in many philosophical issues beginning at an early age. After stepping onto the world stage in 1915, he not only spoke and wrote about the many moral, political, and philosophical issues mentioned above, he also corresponded with the famous Austrian psychiatrist Sigmund Freud about the love-hate instincts of mankind, engaged in extended philosophical dialogue on the nature of truth with the Hindu poet and mystic Rabindranath Tagore, and debated with many other thinkers of his day. He was also a gifted writer in the German language, a lover of music, and most concerned about needless human suffering.

The following excerpts from Einstein's public statements and articles offer a glimpse into the rich body of ideas, ideals, opinions, and views of life espoused by this modest, brilliant, philosophical, and highly idealistic individual.

ON THE MILITARY

In two weeks the sheeplike masses of any country can be worked up by the newspapers into such a state of excited fury that men are prepared to put on uniforms and kill and be killed, for the sake of the sordid ends of a few interested parties. Compulsory military service seems to me the most disgraceful symptom of that deficiency in personal dignity from which civilized mankind is suffering today. (Published in *Mein Weltbild*, 1934)

ON PALESTINE

Jews and Arabs confront each other as opponents . . . This state of affairs is unworthy of both . . . A Privy Council is to be formed to which the Jews and Arabs shall each send four representatives . . . These eight people are to meet once a week. They undertake not to espouse the sectional interests of their profession or nation but conscientiously . . . to aim at the welfare of the whole population of the country . . . (Munich, 1930)

ON RIGHT-WING ANTI-COMMUNISTS

The reactionary politicians have managed to instill suspicion of all intellectual efforts into the public by dangling before their eyes a danger from without. Having succeeded so far, they are now proceeding to suppress the freedom of teaching and to deprive of their positions all those who do not prove submissive . . . (New York, 1953)

ON HUMAN RIGHTS

The existence and the validity of human rights are not written in the stars. The ideals concerning the conduct of men toward each other and the desirable structure of the community have been conceived and taught by enlightened individuals in the course of history. Those ideals and convictions . . . have been trampled upon . . . A large part of history is therefore replete with the struggle for those human rights, an eternal struggle in which a final victory can never be won. But to tire in that struggle would mean the ruin of society. (Chicago, 1954)

The international significance of all the causes on which he spoke out enlarged his human figure to great proportions.

Einstein Looks for Order in a Chaotic World

Traveling on a Beam of Light

Albert Einstein was far more than a scientist who pondered space and time. Yet it is his soaring scientific mind for which we know him. He was born on March 14, 1879, in Ulm, Germany. About a year later, his father's business collapsed and the family moved to Munich with Einstein's uncle Jakob to set up a small electrical plant and engineering works. Einstein attended rigidly disciplined schools in Munich, where he developed his lifelong dislike for regimented education and mindless discipline, as seen in his numerous public criticisms of military authority and totalitarian states.

The stories about his poor scholastic performance in these early years are true. At school, he was bored, intimidated, shy, and withdrawn. Historians have even suggested that he suffered from a form of dyslexia, but that has not been confirmed. Einstein's brother once wrote that Albert's "teachers reported . . . that he was mentally slow, unsociable and adrift forever in his foolish dreams." Responding to the school's headmaster about what profession Albert might enter some day, Einstein's father said, "It doesn't matter . . . he'll never make a success of anything."

There were a number of influences in his early childhood that played a part in setting him on the path to relativity: a pocket compass, given to him at age five, which indicated to him that space was not empty, as he'd been taught; his love for the violin, which led him to an awareness of the mathematical structure of music; his uncle Jakob Einstein, who made mathematics interesting; and his uncle Cäsar Koch, who recognized and supported his interest in science. But it was Max Talmey, a young medical student at Munich University, and Einstein's tutor and mentor, who first sparked his latent brilliance by giving him advanced books on physics and mathematics. Talmey said, "His mathematical genius was so high that I could no longer follow. Thereafter philosophy was often a subject of our conversations. I recommended to him the reading of Kant." Later Einstein also read the philosophical works of David Hume, Ernst Mach and others.

By the time he was 12 years old, he'd already developed a healthy skepticism for much of the material forced upon him in his earlier years in the regimented German school system. He began to focus on the physical world and refine his thoughts about it. Einstein was expelled from school at age 15 for refusing to submit to that regimentation. He was told by his instructor: "Your presence in the class is disruptive and affects the other students." He then proceeded to Milan, Italy, to rejoin his family, who had moved there from Germany because his father's business had failed once more. The next year, at age 16, while Einstein was riding his bicycle through the Italian countryside, he asked himself an innocent question that would lead to one of the greatest scientific discoveries in history—one that fundamentally altered our view of the universe: *What would the world look like if I were sitting on a beam of light, moving at the speed of light?*

Einstein wanted to continue his education in physics and math by obtaining admission to the renowned Polytechnic Academy in Zurich, Switzerland. However, in the spring of 1895 he was only 16 years old, and one had to be at least 18 to take the entrance exam for this school. But that year he wrote a paper about electromagnetism that foreshadowed the theory of relativity, and sent the paper to his uncle Cäsar. Through family friends, the paper made its way to the principal of the academy, and that eventually led to Einstein's admission. His goal was to become a teacher of mathematical physics, so his subjects included differential and integral calculus, descriptive and analytical geometry, the geometry of numbers, and the theory of the definite integral. In these final years of the last century, new theories about electromagnetism emerged, and scientists recognized that Newton's physics could not explain certain phenomena. This new knowledge began to shake the foundations of accepted thought about the universe. During Einstein's four years at the academy, his thoughts were being formed simultaneous with Röntgen's discovery of X-rays, Thomson's discovery of the electron, and Ernest Rutherford's gradual formulation of the modern concept of the atom. Yet Newtonian physics had been accepted, had worked for so long, and was so firmly entrenched that the conservative establishment of science was extremely reluctant to acknowledge any new concept that might reflect any shortcomings in that body of knowledge.

Einstein's ability to conceive new approaches without the shackles of accepted scientific dogma, combined with the influences of his uncles and Max Talmey, the recent discoveries in the field of electromagnetism, and four years of education in physics and math at the Polytechnic Academy set the stage for one of the greatest intellectual advances in the history of science. In the meantime, however, Einstein continued to display his rebelliousness and independence at the academy. "You have one fault," a professor advised him. "One can't tell you anything." As a result, upon graduation he was passed over for a position as an assistant professor in the physics department at the academy, dashing his hopes to become a teacher of mathematical physics.

Einstein Redefines the Universe

An Unknown Clerk . . . The Equivalence of Mass and Energy

Einstein returned to Milan to live with his family again for several months, but soon returned to Zurich and became a Swiss citizen. With prospects of permanent employment now greater than they'd been when he was a German Jew, on June 16, 1902, Einstein was formally appointed as Technical Expert at the Swiss Patent Office in Bern, at a salary of 3,500 francs a year (the equivalent of about

$3,000). He lived in a one-room apartment a few hundred yards from the office. His boyhood mentor Max Talmey visited Einstein in Berne and later wrote, "His environment betrayed a good deal of poverty. He lived in a small, poorly furnished room ... His hardships were aggravated through obstacles laid in his way by people who were jealous of him." Day after day, Einstein sat in the long, narrow room of the patent office, along with his fellow technical officers, reading the endless patent applications for typewriters, cameras, engineering instruments, and the myriad of other assorted devices for which hopeful inventors wished to claim legal protection. He soon mastered the undemanding routine, thereby freeing his creative genius to develop his ideas about physics in his spare hours away from the office.

To earn extra money, he placed an ad in the newspaper offering private tutoring in physics. This led to a following of bright young physics students two to three years his junior. Einstein socialized with his students and engaged in long intellectual discussions and debates that continued over the next several years. They often took trips together through the Swiss countryside and pursued their common passion for physics at Cafe Bollwerk, near the patent office, as Einstein refined his thoughts and improved his ability to express his theories in the most cogent manner possible. Some of these students later wrote that even in his early twenties not only was Einstein's genius apparent, he also had a commanding presence and impressive force of character.

In 1903, Einstein married Mileva Maric, a friend from his days in Zurich. Two years later, still an unknown clerk in the Swiss Patent Office, the 26-year-old Einstein wrote an article, "On the Electrodynamics of Moving Bodies," in which he answered the question he'd asked himself ten years earlier. Based on his mathematical calculations contained in the article, Einstein concluded that if light always moves with the same speed in free space, regardless of the motion of the source, the passage of time must be relative, not absolute. In other words, a person moving with light, at the speed of light, could click his or her own flashlight on, and the beam of the flashlight would leave that person and the flashlight *at the speed of light* relative to that person *and* relative to a non-moving observer. The concepts and proof supporting this conclusion became known as the special theory of relativity (which will be examined in more detail in Chapter 8) and resulted in a fundamental change in how we view the universe. In a related paper Einstein established the equivalence of mass and energy, later represented by the famous equation $E=mc^2$, which was examined in Chapter 6. This was the most important conclusion that Einstein derived from the special theory of relativity.

In 1905 he published a total of five papers on entirely different subjects. Three of them were among the greatest in the history of physics. The paper concerning the quantum explanation of the photoelectric effect earned him the Nobel Prize sixteen years later. The third paper contained the special theory of relativity, adding space-time as the fourth dimension of the universe—nine thousand words that comprise one of the most remarkable scientific papers ever written.

Nevertheless, it took time for other physicists to appreciate the significance of Einstein's papers. In 1907 Einstein took the first step in what would become a long academic career when he became an instructor on the theoretical physics faculty at the University of Bern, while retaining his job at the patent office. There were only four students in his first class and one in the following term. As relativity slowly percolated through the scientific community and others began to appreciate what this new body of knowledge represented, Einstein was drawn out of his limited and isolated shell and gradually embraced as a full member of the international scientific community. He began spending more and more time away from Bern, as he established relationships with Hendrik Lorentz in Holland, Ernst Mach in Austria, Ernest Rutherford in England, Marie Curie and Paul Langevin in France, Max Planck in Berlin, and Arnold Sommerfeld in Born. By 1909 he was receiving numerous invitations to lecture at various scientific conferences, and was soon regarded as a leader of the scientific elite of the early 20th century that was focused on discovering the nature of the physical world.

On July 6, 1909, Einstein resigned from the patent office and moved back to Zurich to take up his first full-time academic position in the newly established chair of theoretical physics at the University of Zurich. Einstein's lectures at Zurich (on thermodynamics, the kinetic theory of heat, electromagnetism, and other topics in physics) became extremely popular, due to his humor, unusual presentations, patience, and accommodation to his students to make sure they understood.

Those next few years were dominated by his continued travel throughout Europe and by the refinement and consolidation of his scientific theories, including development of his general theory of relativity (which will also be discussed in Chapter 8). In March 1911 he accepted a position at the German University in Prague, Czechoslovakia, a few miles from the place where Tycho Brahe first employed the young Johannes Kepler. He noticed a growing hostility between the Czech and German citizens there. Unbeknownst to him and to the world, this animosity foreshadowed an increasingly volatile political situation in Prague that would precipitate world war.

By this time the Einsteins had two children—Hans, born in 1903, and Edward, born in 1910. In March 1912 Einstein went to Paris to address the French Society of Physicists, and strengthened his friendship with Marie Curie, which would continue until her death in 1934. In July of that year, their families took a joint vacation in the Alps. The Curies' eldest daughter, Irène, eventually took up the work of her mother and also established a long friendship with Einstein.

In April 1914 the family moved to Berlin, where Einstein began working with the generous support of Germany's premier science association, the Prussian Academy of Sciences. Einstein expressed his gratitude in this letter:

> Gentlemen . . . I have to thank you most heartily for conferring the greatest benefit on me that anybody can confer on a man like myself. By electing me to your Academy you have freed me from the distractions and cares of a professional life and so made it possible for me to devote myself entirely to scientific studies.

These words and the mutual admiration between Einstein and

the academy would soon be swallowed up in the darkness that engulfed the world as the Great World War erupted. The changing face of the world and Germany's pivotal role in it immediately changed Einstein's life and led to his transformation from a quiet scientist to a champion of pacifism.

Because of the war, Mileva and their two sons were not able to return to Berlin after their summer vacation in Switzerland in 1914. This involuntary separation eventually led to divorce. In 1915 Einstein began to speak out against German militarism and nationalism in general. He focused his life on two matters—the World War and finalizing his *general* theory of relativity. He first published that theory in 1916 in an article entitled, "The Foundation of the General Theory of Relativity." Gravitation, Einstein explained, is not a force, despite what Newton had thought. Instead, "it is a curved field in the space-time continuum." Einstein's excitement over concluding this work was revealed in a letter to his friend, the German physicist Arnold Sommerfeld (1868–1951) in November 1916, when he said, "This last month I have lived through the most exciting and most exacting period of my life, and it would be true to say that it has also been the most fruitful period." Other physicists recognized the combined theories of relativity as a revolution in thought, the likes of which had not been seen since Newton's *Principia*.

Einstein Steps Onto the World Stage

Hitler's Greatest Public Enemy

In 1919 Einstein married Elsa Lowenthal, a distant relative of his late father, and lived with her and her two daughters in Berlin. In that same year, the Royal Society of London announced that, based on its photographs of the solar eclipse that occurred on May 29, 1919, it had completed calculations verifying the predictions that Einstein had made in his general theory of relativity concerning the deflection of starlight. Although Einstein had already become well known in scientific circles throughout the world, true international fame came to him as a result of the Royal Society's announcement.

His public life changed dramatically. He became the symbol of science and the master of 20th-century intellect. For the next thirty-five years, he became outspoken on the long list of social causes mentioned earlier, primarily centered around his promotion of international pacifism. Through the 1920's, a period of unrealistic ideals, Einstein embodied idealism as he served on a number of organizations that promoted his philosophies, such as the newly formed League of Nations International Committee on Intellectual Cooperation, along with Marie Curie and other physicists.

By 1929 Einstein had established strong ties with physicists all over the world, including in the United States. With the resurgence of German militarism, accompanied by an unprecedented and most sinister level of anti-Semitism, Einstein began to think of not only leaving Germany, but fully detaching himself from the entire European continent. By mid-1932, when it was obvious to Einstein that the Nazis would soon gain power in Germany, Einstein and his wife were already making plans to move to the United States. While he was on a lecture tour there, Einstein's fate was sealed when on January 30, 1933, Adolph Hitler came to power as Chancellor of Germany. Einstein then renounced his German citizenship, condemned the country of his birth, and urged all of Europe to take up arms against the Nazis and prevent them from precipitating a war. His predictions in these early years of Hitler's reign were ignored. In retaliation for Einstein's public statements, Nazi storm troopers ransacked his summer house outside of Berlin. Einstein's relationship with the Prussian Academy of Sciences, which had embraced him in 1914, abruptly ended when Einstein publicly withdrew and the academy issued its "Declaration of April 1, 1933," in which it vowed allegiance to Germany and stated that it had "no reason to regret Einstein's withdrawal."

In October 1933 he accepted a full-time position at the school of mathematics in the newly founded Institute for Advanced Study at Princeton University, in New Jersey, where he would remain for the following twenty-two years. As if the special and general theories of relativity had not been great enough accomplishments, Einstein pursued the question of whether it would be possible to unite those and other theories into what would be called the "unified field theory." This scientific pursuit would dominate his remaining years but would never bear fruit.

His scientific and social pursuits were not nearly as influential or significant after the 1920's as they'd been up to that time. In addition, his son Edward suffered a mental breakdown and blamed his father for deserting him after the 1914 divorce from Mileva, and for being the cause of his troubled life. After this crushing development and after the death of his wife Elsa in 1936, Einstein became increasingly more solemn and discouraged by his own life, including his inability to slow the threat of war that he knew would be disastrous to so many innocent people. As seen in Chapter 6, his quiet work at Princeton was interrupted in 1939 when he agreed to sign the letter to Roosevelt, which led to the Manhattan Project. Otherwise, his travels lessened through the 1940's, and he maintained a routine existence centered on his theoretical research, but didn't make any further meaningful contribution to physics.

Measured by his lasting contribution to our understanding of the universe, Albert Einstein will always remain a great figure in science. But to fully appreciate him as a human being, one must also understand his other passions—peace, the human condition, truth, and the difficult task of creating political institutions that maintain freedom in society. Einstein reprimanded the failed governments and the oppressive policies that determined millions of people's fates during the first half of this century. He became the conscience of the world. But those crusades failed. Ironically, his only effort that resulted in the establishment of a clear government policy was the fateful letter to Roosevelt encouraging the establishment of the research that led to the Manhattan Project and forever linking Einstein's name to the production of the atom bombs that killed over 200,000 people and destroyed Hiroshima and Nagasaki in August 1945. Einstein expressed regret about this in many different ways at different times. For example, he said to the famous biochemist Linus Pauling, "I made one great mistake in my life—when I signed the letter to President Roosevelt recommending that atom bombs be made." He also wrote, "The physicists who participated in forging the most formidable and dangerous weapon of all times are harassed by a feeling of responsibility, not to say guilt . . ." He saw himself and those other physicists as collaborators in the invention of the worst death machine in history. Yet, in the context of the time when he made the decision to sign the letter, it was a logical decision, one that he sometimes defended:

> We helped in creating the new weapon in order to prevent the enemies of mankind from achieving it ahead of us, which, given the mentality of the Nazis, would have meant inconceivable destruction and the enslavement of the rest of the world. We delivered this weapon into the hand of the American and British people as trustees of the whole of mankind, as fighters for peace and liberty. But so far we fail to see any guaranty of peace. We do not see any guarantee of the freedoms that were promised . . . The war is won, but the peace is not.

Thus, based on the possibility that the Germans would develop the bomb first, he felt the Manhattan Project was justified. But when the United States later discovered that the Germans were not as far along as we feared, and when the war in Europe ended in May 1945, he opposed using the weapon. "I have always condemned the use of the atomic bomb against Japan," he said. Einstein's dedication to pacifism, which became an even greater passion after the bombs were dropped on Hiroshima and Nagasaki, centered on the belief there should be one world government under a constitution to be jointly drafted by the United States, Britain, and Russia. "I advocate world government," wrote Einstein, "because I am convinced there is no other possible way of eliminating the most terrible danger in which man has ever found himself."

Einstein's position on this point was made without regard to the powerful social, cultural, and political forces that bind groups into national and regional entities. These forces have roots that go back hundreds or even thousands of years, and they influence events regularly around the world—in places like Bosnia, Chechnya, the Middle East, Somalia, Rwanda, Haiti, and Northern Ireland. Einstein must have known that those forces cannot be subdued by a vague and untested prediction that the greatest good for the greatest number of people will result if nations universally relinquish their sovereignty.

Although Einstein's social and political views never had a significant impact on those who controlled governments, he remained a respected figure through most of his life. However, after World War II, his pacifist and one-world-government views, in a nation increasingly opposed to and distrustful of the Soviet Union, led to a broad distrust of Einstein himself. He suffered a final though gentle fall from grace after his address to a television audience in 1950:

And now the public is advised that the production of the hydrogen bomb is the new goal . . . An accelerated development towards this end has been solemnly proclaimed by the President. If these efforts should prove successful, radioactive poisoning of the atmosphere, and, hence, annihilation of all life on earth, will have been brought within the range of what is technically possible.

In his waning years following this speech, he persisted in his call for pacifism, yet was viewed as increasingly out of context, out of his element, and as a lone, eccentric scientist out of touch with social and political reality—a view largely supported by his description of himself:

My passionate interest in social justice and social responsibility has always stood in curious contrast to a marked lack of desire for direct association with men and women . . . I have never belonged wholeheartedly to any country or state, to my circle of friends, or even to my own family. These ties have always been accompanied by a vague aloofness, and the wish to withdraw into myself increases with the years. Such isolation is sometimes bitter, but I do not regret being cut off from the understanding and sympathy of other men. I lose something by it, to be sure, but I am compensated for it in being rendered independent of the customs, opinions, and prejudices of others . . . It is strange to be known so universally and yet to be so lonely.

Albert Einstein had no illusions about whether his words could sway nations. But he was morally compelled to speak and write those words. He wanted "simply to serve with my feeble capacity truth and justice at the risk of pleasing no one." Through his devotion to the activities and state of humanity, and as a person profoundly concerned with science as a part of our culture, Einstein—the German who hated Germany, the Zionist who tried to make peace with the Arabs, the pacifist who had a hand in the birth of nuclear weapons, and the introverted loner who became a spokesman on the world stage—demonstrated better than anyone how dependent science is on a free society, and how fleeting that freedom can be.

Chapter Eight

The Fourth Dimension

Sometimes I see
clearest of all
when I'm alone and remember
afternoon sunlight at schoolday's end
the world through the eyes of a child.

Then I can see
a world yet to be
here and now in my grasp,
like Saturday mornings when I wake and see
the world through the eyes of a child.

When I grow old
in Winter and twilight
I will remember Spring
and wisdom I learned
from that sometime view of
the world through the eyes of a child.

JOHN KUZMA
The World Through the Eyes of a Child (1996)

Everyone knows that Einstein was a scientific genius responsible for doing something extraordinary, but few people know what it was. This chapter will attempt to explain relativity. To *fully* understand it, one needs knowledge of advanced physics, but like the six other great scientific discoveries covered in this book, relativity can be grasped conceptually and without equations. Yet, relativity does require a higher level of abstract thinking than is necessary to understand the other discoveries. We don't experience relativity like we do gravity, for example, and we can't *see* relativity like we do the galaxies or the human cell. Einstein himself said that the advantage of most theories is their clarity, while the advantage of proving and understanding relativity is its "logical perfection." The principle of relativity is the one most at odds with common sense, so it requires looking at the world unfettered by a lifetime of experiences. It requires seeing the world through the eyes of a child.

Newtonian Physics Falls Short

Questions About Mercury's Orbit

We will first examine Einstein's special theory of relativity. At the end of this chapter we'll briefly look at his general theory of relativity. *The central significance of the special theory of relativity is that the passage of time is not absolute.* "Absolute, true and mathematical time," Newton wrote in the *Principia*, "of itself and from its own nature, flows equably, without relation to anything external . . ." Remarkably, numerous experiments testing the special theory prove that Newton's statement is not true. The rate at which time passes is different for one person "at rest" than it is for another who is *moving* at great speed in relation to the person at rest, even though for both of those people *their* time is the real, actual, and normal time. With the discovery of relativity, we realized space and time are not separate. They are intertwined as *space-time*. There was nothing wrong with Newton's physics, but it had to be extended to explain the relationship between velocity and space-time. Special relativity didn't overthrow Newtonian physics, it only demonstrated that Newton's principles were not valid in extreme conditions. When Newton referred to distance, he meant distance at a given time. No one, including Newton, thought there could be any ambiguity about "time."

The laws and principles discovered by Newton and revealed in the *Principia* in 1687 had provided all the answers necessary for physics to operate successfully for over 200 years. Galileo and Newton studied "relativity" as that term was used in their time. Some of the ideas they developed were an elaboration of concepts known since ancient Greece, while others were a correction of ancient beliefs. In all cases, Galileo's and Newton's laws on this subject were universally accepted before Einstein's theory and are known as "Galilean relativity." Here are two examples of Galilean relativity:

1. If a person is walking up the aisle of a train, in the same direction as the train, his or her speed *relative* to the ground is the sum of the train's speed and the walking speed. For example, if the train is moving at seventy miles an hour, and the person is walking at four miles an hour, the person's speed *in relation to the Earth* is seventy-four miles an hour.
2. If a person on a moving train drops an object from the ceiling to the floor, it will drop straight down relative to the train and to the person on the train. However, relative to Earth and as seen by an observer who is standing ("at rest") outside the train, it will appear to drop in a motion that looks like a *curve*, since the vertical velocity of any falling object increases with time. This curve is shown in Figure 8-1.

The above examples of Galilean relativity show that some type of *frame of reference* (such as the train, the Earth, another planet or other heavenly body) is needed to describe the movement of an object.

Toward the end of the 1800's, facts were slowly being uncovered that seemed to conflict with or were not addressed by the prevailing laws of physics, including Galilean relativity. For example:

a. The speed of light did not behave like speeds of material objects;
b. The heat capacity of molecular gases differed from values calculated using Newtonian theory;

Figure 8-1

GALILEAN RELATIVITY—OBJECT DROPPED IN MOVING TRAIN

Five Successive Positions of Train

Direction of Train and Object ⟶

```
0                    50                   100
              (Feet of Travel)
```

c. Radioactivity showed that matter was subject to unpredictable eruptions of instability;

d. Mercury's orbit was at variance with Newton's laws of motion; and

e. There was no explanation in classical physics for the existence of light and other radiation emitted by atoms.

For the first time since Newton, scientists began to question the validity of the classical framework in which they had put so much faith. Their instruments were becoming more refined and sophisticated, thus enabling them to identify these ever so slight discrepancies. One discrepancy in particular—the mystery that surrounded point (a), above, regarding the speed of light—became explainable when Albert Einstein conceived his special theory of relativity in 1905. Again, the central significance and the greatness of the special theory is the principle that *the passage of time is not absolute.* The foundation of the theory is based simply on Einstein's combination of two other principles: Galilean relativity plus a principle that he discovered called the constant velocity of light. This principle holds that light always has the same velocity in a vacuum (186,000 miles per second), *regardless of the motion of the observer or the source of the light.*

Einstein's discovery of the principle of the constant velocity of light was the direct result of the question that Einstein had asked himself at age 16. In his autobiographical notes later in life, he explained why he became so certain of the principle of the constant velocity of light:

> After ten years of reflection such a principle resulted from a paradox upon which I had already hit at the age of sixteen: If I pursue a beam of light with the velocity [of light], I should observe such a beam as a spatially oscillatory electromagnetic field *at rest.* However, there seems to be no such thing . . . From the very beginning it appeared to me intuitively clear that . . . everything would have to happen according to the same laws as for an observer who, relative to the earth, was at rest. For how, otherwise, should the first observer . . . be able to determine that he is in a state of fast uniform motion?

The Scottish physicist James Clerk Maxwell (1831–1879) had published his two famous papers in 1861 and 1865 on the existence of an invisible electromagnetic field that included a broad range of *waves* (X-rays, radio waves, light waves, etc.). In other words, Maxwell had shown that light waves are simply the *visible* portion of this spectrum and that all these waves travel at 186,000 miles a second. Einstein theorized that the speed of these electromagnetic waves (including light) must be a constant and not a relative quantity. The speed of the observer shouldn't make a difference in the speed of such waves. Either Maxwell's equations were incorrect or Newtonian mechanics was incorrect. The need to resolve this inherent conflict led Einstein to the special theory, which asserts that Maxwell's equations are correct and Newtonian physics is not adequate to explain space and time when an object or observer approaches the speed of light.

The Michelson-Morley Experiment Is Resurrected to Support Relativity

Vibrations in the Ether

The constant velocity of light was actually demonstrated twenty years before Einstein stated it as a principle. From 1881 to 1887, the German-born American physicist Albert A. Michelson (1852–1931) conducted a series of experiments using light, as described below. This work culminated in a final and now famous experiment in 1887 by Michelson and the American chemist Edward W. Morley (1838–1923), which became known as the Michelson-Morley experiment. Although these experiments were aimed at proving a different theory and were not correctly interpreted as applying to the speed of light until Einstein came forth with the special theory of relativity, they later provided the best data available to support Einstein's postulate on the constant velocity of light, which is critical to understanding relativity.

Even after Maxwell's discovery of electromagnetic radiation, most physicists believed that space was filled with an invisible weightless substance that actually oscillated, in wave form, in order for light to propagate, similar to sound resulting from the vibration of air. The supposed substance was called "luminiferous ether" (or just "ether"). Despite Maxwell showing that light itself is an electromagnetic phenomenon—that is, waves and not particles—most physicists continued to believe in some type of ether to "support" or conduct the light waves.

The early experiments by Michelson and the Michelson-Morley experiment of 1887 were intended to confirm the existence of the ether. They tried to prove this by measuring the difference in the time it takes light to travel in the direction of the Earth's motion in its orbit around the Sun compared to ninety degrees (or transversely) to that direction. In the framework of Galilean relativity, before it was known that the speed of light is constant, one would expect there to be a difference, since—like the person walking up the aisle of a train or the object being dropped on a train—a beam of light directed *toward* the Earth's direction of motion around the Sun would be expected to travel at the speed of light *minus* that orbital movement of about 67,000 miles an hour, like a boat moving upstream in a river. It could then be compared to the speed of light traveling in other directions, and from that data the speed of the Earth relative to the ether—the object of the experiments—could be calculated. In 1887, Michelson and Morley (both working in Cleveland at the time) repeated the earlier experiments by Michelson and produced the same result, namely, *no detection of the Earth moving through the ether.*

If one forgets about the ether theory, as Michelson, Morley, and other scientists should have done, the experiments showed that the speed of light was the same in all directions at all times. That is, light doesn't change its velocity depending on whether it's traveling "upstream" or "downstream" through the nonexistent ether. With modern technology, the principle of the constant velocity of light has been confirmed beyond doubt.

In 1907, Michelson became the first American ever to receive the Nobel Prize in the sciences, for his spectroscopic and metrological (science of measurement) discoveries. Even though Michelson's work is now regarded as very significant in the history of physics, he felt that the experiments were a total failure because he never achieved his objective of measuring the speed of the Earth relative to the ether.

Einstein Determines That if Light Speed Is Absolute, Time Is Not

Moving Frames of Reference

Given the principle of the constant velocity of light, we'll now return to special relativity, remembering its central significance: *the passage of time is not absolute.* The special theory only pertains to the relationship between two objects, where one is moving toward or away from the other at great speed—that is, the movement of one relative to the other—thus, the name for the theory. Obviously, there would be no *relativity* in a universe with only *one* object in it because there's nothing to be relative *to*. But in our universe, if one object or observer is at rest *in relation* to another object or observer that is moving at great speed, that second object's or observer's *movement in relation to the first* results in a difference in the passage of time for each observer, as measured by the other. If observers A and B are at rest in relation to each other, but C is moving at great speed relative to A and B, then the passage of time for C will be measured by A and B to be slower.

The special theory of relativity began in Einstein's mind as a concept. He then created the mathematical proof to support it. Numerous experiments in the 20th century have provided the supporting empirical evidence. We will not delve into any of the math or the experiments, but we can gain a clearer conceptual understanding if you refer to Figure 8-2 and consider A, B, and C as *objects or observers*, with A and B in one frame of reference—that is, a physical area in space or a body where A and B are stable and not moving *in relation to each other*—while object or observer C is in another frame of reference.

There are actually three ways to look at the A-B-C relationship when the space between the pair A-B is increasing relative to C: (1) C is moving away from A-B; (2) A and B are moving away from C;

Figure 8-2

A-B FRAME OF REFERENCE RELATIVE TO C

C is moving away from the A-B frame of reference

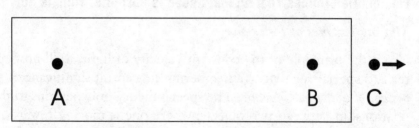

A is at rest relative to B   B is at rest relative to A or (3) A-B and C are moving away from *each other*. We will examine this as if (1) were occurring—that is, A and B are "at rest" and C is moving as shown in Figure 8-2. We'll use this as a model in this chapter so we have a consistent framework and viewpoint, even though (2) and (3) are equally valid.

Light moving between A and B in *one* frame of reference covers a certain distance in a certain period of time.

Figure 8-3

LIGHT WITHIN THE A-B FRAME OF REFERENCE

A   B

Beam of Light

However, what if C is rapidly moving away from the A-B reference frame, as suggested in Figure 8-2? Given that the speed of light does not change, regardless of the movement of the body from which the light originates, how do we reconcile the speed of a light beam flashed from A to B (that is, *within* one frame of reference) with what you would expect to be a *slower* speed of light flashed from the A-B reference frame *toward* C, as measured in frame C? That is, you'd expect it to be slower because if Galilean relativity were applied to the two rapidly separating frames of reference, a "net" speed of light would result by subtracting the speed of the separation from the speed of light.

But we are given the principle of the constant velocity of light, and it must *somehow* be reconcilable with the relative movements of the two frames of reference. The special theory of relativity succeeded in this reconciliation by proving that time is "personal" to observers in separate inertial systems. That is, since light's actual

Figure 8-4

LIGHT TRAVELING FROM A-B TO C

speed over a distance is constant, the time element must *not* be constant. It must depend on *the observer's relationship to the moving object*. We are all in the same frame of reference on Earth and traveling slowly—therefore, passage of time is constant and does not change. But if we were in reference frame A-B, and C was moving toward us at the speed of light, and if C flashed a beam of light toward A-B, light would *not* be traveling toward A-B at *twice* its normal speed (that is, 372,000 miles a second). Travel speed and light speed can't be *added* together because the velocity of light is constant.

The theory can be further understood by answering and examining these three questions:

1. How fast do you have to be going before your speed has an impact on how fast time passes relative to someone "at rest"? If there is one clock in the A-B frame of reference and another clock in frame C, how fast does C have to go before there's a significant difference in the rates of the two clocks as measured from either frame?
2. How can a person visualize and gain a clear conceptual understanding of this new perception of time?
3. What happens to matter physically when it's moving at these great speeds?

Once you understand the answers to these three questions, you will understand special relativity on a conceptual basis.

Question Number 1
What Is the Relationship Between Speed and Time?

The Relativistic Factor

Remarkably, Einstein's theory would give a different result of the speed of the person walking down the aisle of the train than would Galilean relativity, though the tiny difference would be extremely difficult to measure. But if we are dealing with observers and objects under observation that are moving *relative to each other* at velocities over about half the speed of light, Einstein's theory is the only correct way to determine the result. This is shown in Figure 8-5.

The commonly observed phenomena in our daily lives are adequately described by Galilean relativity and Newtonian physics. However, this graph shows that once any matter approaches the velocity of half the speed of light, the effect on time becomes in-

Figure 8-5

Correlation Between Speed and Time Dilation

creasingly dramatic because there is an increasing and much more noticeable effect of speed on time "intervals"—that is, the time *between* events, the very rate at which time passes. So, for example, the time dilation at half the speed of light is noted on Figure 8-5, and shows that if an object or observer (such as C) is moving at that speed, time intervals in C's frame of reference will be 15 percent shorter than the time intervals between the same events as measured in frame A-B, which is the frame of reference at rest. One hour for a person moving at half the speed of light is equivalent to one hour and nine minutes for a person on Earth, because 115 percent of 60 minutes is 69 minutes. This ratio between the two time periods is calculated by multiplying C's time by Einstein's "relativistic factor" (which is derived by an equation he developed). A space traveler's life will actually be extended relative to an Earthbound person, even though the passage of time for the person traveling at half light speed and everyone with him or her will *feel* no different because that will be their *actual* time. It is not a *perceived* change in time. It is their *actual* time. Moving clocks run slower.

The idea that time can actually change and is a function of velocity is totally contrary to our everyday experience. Time seems so firm, so constant, and so "un-relative" that any tinkering with it is difficult to accept. Under Newtonian physics, *mass, length* of any physical object, and *time* were the sole fundamental quantities needed to describe mechanical movement of such an object. Einstein added his "relativistic factor" and showed that as such movement approaches the speed of light, the time that *actually* (not theoretically) passes in the moving frame will be less, relative to the actual time that passes in a frame of reference at rest.

The purpose of the relativistic factor is to *adjust* or *correct* for the two time frames. Without such an adjustment, we would be left with the irreconcilable results that do not account for the principle of the constant velocity of light. Indeed, Einstein's principle would be more accurately called "Absolutism" or "The Special Theory of *Non*-Relativity" because he was seeking a *unifying* principle and the *removal* of the relative nature of the fundamental laws of Newtonian physics. He did not name his principle the special theory of relativity—that was a term invented by others.

One more example will further illustrate the relationship between speed and time and that the passage of time is not absolute. Imagine that in the A-B reference frame, with A and B at a great but constant distance from each other, A aims his flashlight at B and turns it on for an instant once every two minutes, sending a *white* beam of light to B. Thus, B sees the beginning of each successive flash every two minutes.

Figure 8-6

Two-Minute Intervals Between Flashes

Since A and B are in a constant position relative to each other, the first instant of each flash takes the same amount of time (two minutes) to reach B. We don't know the time it takes for the *first* flash of light *to travel from A to B*. It could be a minute or ten years, but it doesn't matter because all we are concerned with is the two-minute *interval between* the *beginning* of one flash and the *beginning* of the next.

Add to this example the assumptions that person C is traveling *away from* A toward B at a constant speed and while C is moving she's watching A's flashes (See Figure 8-7). Of course, each flash reaches C before it reaches B, since she's closer to A, and—even at a very slow speed—since C is moving *away* from A, it takes *more* than two minutes *between* flashes.

In other words, if C had stopped at the exact location where she was when the first flash reached her, the second flash would have come two minutes later—the same *interval* as A's actual flash and B's

Figure 8-7

C Traveling from A to B

receipt of it. But since C is moving away from A and each succeeding flash has to travel farther than the previous one to reach C, the second flash would take more than two minutes to get there. This would happen without regard to any relativity principles—Galilean or Einsteinian. The interesting phenomenon is not what happens between A and C, but rather the effect of special relativity between C and B, as will be described below.

Now add these assumptions to this example: C's speed is 75 percent of the speed of light, and C is carrying a light source of her own and is sending a second beam of light, let's say *yellow*, to B beginning at the precise moment that A's white beam gets to C. (See Figure 8-8.) Although C is moving, her yellow flash would still

Figure 8-8

C Traveling from A to B and Sending Beam to B

leave at light speed and travel to B at light speed. Thus, will C's yellow beam reach B *before* A's white beam?

No, as discussed earlier, that would be impossible. The two beams would reach B *at the same time* because they are leaving C and are traveling together from C to B *at the same time*, even though they emanate from sources that are not in the same frame of reference—white beam from A and yellow beam from C.

Applying Einstein's relativistic factor shown in Figure 8-5, at 75 percent of the speed of light, the same two-minute interval for A and B would increase 50 percent for C. It would be a *three-minute* interval for C. How can the two beams reach B together when the white beams or flashes (A to B) reach B every *two* minutes in the A-B frame of reference, while the yellow beams or flashes (C to B) reach B every *three* minutes? The only answer is that "time" for A and B is not the same "time" for C. *Since the white and yellow beams of light require two different time periods in order to cover the same distance and in order to be explained to two separate observers, then the time component must not be constant.*

As long as A and B are in a constant (unmoving or inertial) relationship *to each other,* their time intervals will match. But their perception of elapsed time will never be the same as that of an observer moving (at great speed) toward or away from them. This is the only possible explanation for the fact that B sees A's white beam arriving at B the same time as C's yellow beam, even though C's yellow beam has the "advantage" of C's *additional* speed toward B. From B's perspective as the receiver of both beams, they look exactly the same—at the instant the white beam reaches and then "leaves" C, the yellow beam travels with it. *They must reach B together every two minutes.* Yet from C's perspective, they are leaving her every *three* minutes. The time *interval* has increased 50 percent in C's reference frame. The rate of the passage of time for C is slower than for A and B.

Returning to Galilean relativity for a moment, if it were a white *ball* being thrown from A to B instead of a white beam of light, with C launching her own yellow ball toward B at the moment the white ball gets to her, of course, C's speed would be used in calculating how fast she would have to throw the yellow ball to cause it to keep astride the white ball as it goes the remaining distance. So, for example, if C were going 20 miles per hour and the white ball was going 50 miles per hour, C could throw the yellow ball 30 miles per hour, resulting in both balls moving forward together for the remaining distance to B. But remember that with special relativity we have removed the speed variable from the equation (light speed is constant) and removed the ability *to combine the speed of the source* from which the light originates.

Question Number 2
How Can This Concept of Time Be Visualized?

Banking Time in Canadian and American Dollars

To conceptualize time dilation, we can compare the two frames of reference to the exchange rate of two different countries. As all foreign travelers experience, on any given day the currency of one nation is equivalent to a fixed amount of currency of each other nation in the world. Imagine that all Earth-bound people "at rest" (such as A and B) are American or British. Then assume that C is a person in space traveling at half the speed of light and that all such space travelers are Canadian. Further assume that all commodities for sale (including the precious commodity of time) are priced the same in both countries. In other words, one hour of time can be purchased by Americans for the same number of U.S. dollars as Canadians would need in Canadian dollars. In this hypothetical situation, the following exchange rates between the countries apply:

- The American (U.S.) dollar is equivalent to $.87 in Canadian money. In other words, it would only take $.87 for C, a Canadian (the space traveler), to purchase an American dollar or $1.00 worth of American goods.
- Reversing the ratio, A and B, who are American or British (at rest on Earth), would have to pay $1.15 (American) to purchase one Canadian dollar or $1.00 worth of Canadian goods. In other words, an American would have to use more dollars than a Canadian to purchase the same item.

This comparison can be illustrated in the following figure:

Figure 8-9

FOREIGN EXCHANGE RATE COMPARED TO TIME DILATION

A • 0   10¢   20¢   30¢   40¢   50¢   60¢   70¢   80¢   90¢ $1.00 • B
American Dollar

A • 0   10¢   20¢   30¢   40¢   50¢   60¢   70¢   80¢ 87¢ • B
Canadian Dollar

If we now substitute the concept of time in place of the dollars, we see that C, the space traveler, and other Canadians are able to buy more "time" than A and B, the Americans and British on Earth—that is, the Canadians (the space travelers) have the more valuable currency. They have greater purchasing power. Converting the 15 percent greater amount of money into a 15 percent greater amount of time (115 percent of 60 minutes is 69 minutes), it takes C less time (fewer dollars) to traverse the distance from A to B because her time (her dollar) is worth more in relation to that of the Americans and British. Thus, while it would take an American or Brit 100 years, for example, to traverse the distance between A and B, it would only take a Canadian (C) about 87 years to go that same distance (as measured by A or B), because her actual time has slowed relative to the Americans' and British residents' actual time. Just like C would only need $.87 to purchase the $1.00 item, she would need less time (measured by anyone in the A-B frame of reference) to traverse the same distance.

Thus, in effect, the Canadian space traveler (C) is banking "spare time" just like she would be banking spare dollars in connection with the exchange rate between the two frames of reference (Canada and the United States). By banking this time, C's life as a space voyager would be extended relative to the life spans of A and B, who remained on Earth.

Question Number 3
What Happens to Matter Physically at These Great Speeds?

Watching Orbiting Electrons

The answer to Question number 3 is *nothing*. However, even though nothing *happens* to such matter, if we examine the *physical characteristics* of matter itself when it moves at great speed, the concept of relativity becomes even clearer. We'll relate the theory to the physical aspects of matter in three different ways:

a. The correlation between all movement and time dilation, including the actual movement of subatomic particles.
b. The "foreshortening" (along the direction of motion) of matter traveling at great speed, as viewed by an observer in a different frame of reference.
c. Einstein's equation $E=mc^2$, which shows that the mass of an object is equivalent to energy.

A. Movement of Subatomic Particles

Imagine any object or particle of matter on the Canadian's spaceship, such as a hydrogen atom with its electron orbiting the nucleus at a given speed, and further assume an identical hydrogen atom in the frame of reference of the American or British person stuck on Earth. If each person had equipment that could simultaneously observe or otherwise record and measure the number of orbits that each of the electrons made around those two hydrogen nuclei during a period of time, *the actual specific number of orbits would be different for the two atoms.* The observers in the A-B frame of reference at rest would conclude that the electron revolving around the atom on the spaceship is *slower* relative to the electron in the atom on Earth—in the same ratio that could be calculated with the relativistic factor used under Question Number 1 (correlation between speed and time dilation), and consistent with the time dilation concept discussed under Question Number 2 (Canadian/American dollars). Both people could "see" this relative difference because their equipment would agree on the measurements. Of course,

Einstein did not envision this hypothetical, because the structure of the atom was not known in 1905. However, it is an inescapable conclusion.

B. Foreshortening

The second point about the physical effect of moving at great speed is the "foreshortening" of matter in the direction of motion. As we explored in Part Two, matter is composed of atoms whose structure is kept in shape by electrical forces. In the modern concept of the atom's structure, the electrons are considered to vibrate in *waves* around the nucleus, rather than as discrete *particles*. Each atom has a certain radius between the electron waves or cloud and the nucleus, which accounts for the specific distance between atoms. If the distance between them changes, the overall length and shape of the object changes.

Thus, the length of a one-foot ruler is determined by such average radii of the atoms of which it's composed. To an observer measuring that ruler as it goes by at a speed approaching the speed of light (but not to a person traveling with the ruler), the distance between those atoms and therefore the ruler itself will appear *foreshortened in the direction of its motion*. Thus, remarkably, even the shape (including the length) of measuring devices depends on their state of motion. Even the distance *measured* between points A and B in the above example (Figure 8-8) involving the white and yellow beams of light would depend on whether that distance was being measured by A or B on the one hand or by C, the person traveling at great speed relative to them.

Observers in relative motion assign different numbers to length and time intervals between a pair of events, rather than finding these numbers to be absolute. Mass, length, and time intervals are all affected when measured from a moving frame of reference. Einstein's relativistic factor depends on the speed of an object—and mass, length, and time depend on the relativistic factor. According to Newton, of course, there was no "correction," because mass, length, and time were believed to be independent of speed. But in reality the entire geometry of space and the rate at which time passes are both determined by the speed at which matter moves through space. Again, the rate of time passage is not an absolute phenomenon—instead, it is the rate of physical change in the universe, as compared between two frames of reference in constant speed toward or away from each other. Such change includes the movement of hands on a clock, the flow of a stream, or any other physical movements that could be used to define the concept of "time." This further demonstrates that relativity involves a *physical* reality, not a *mental* perception. It applies without regard to minds or senses.

C. Mass Converted to Energy

In the course of developing the special theory, Einstein discovered the most famous equation in science: $E=mc^2$. This equation reflects Einstein's realization that inert mass is simply latent energy. The equation combines the principle of conservation of mass with the principle of the conservation of energy, as examined in Chapter 6 in connection with the development of atomic energy and the bomb. Never before realized or recognized, mass and energy are simply and literally two sides of the same equation. All mass carries energy, and all energy contributes to mass. Energy in the form of light and other electromagnetic radiation (called photons) is not matter because it's not made of atoms. But even photons behave like they have mass. The energy associated with even a small amount of mass is enormous, as we realized through the release of power in the atom's nucleus.

The atom's *mass* (thus, matter in general) increases by the relativistic factor as its velocity approaches the speed of light. Objects moving at great speed have substantial energy of motion (kinetic energy), which gives them this additional mass. Again, energy is convertible into matter. While mass was constant in Newtonian physics, it becomes velocity-dependent in relativity—thus requiring that mass be multiplied by the relativistic factor to obtain an accurate value of mass moving at great speed. In other words, the relativistic factor *equalizes* the two masses and "corrects" the relationship between the objects being compared. This *physical* difference between the matter at rest versus the matter moving at great speed is explained by $E=mc^2$ in combination with the relativistic factor, and is part of the very nature of matter. Referring back to the Canadian and American dollars in Question Number 2, the "mass" of the American dollar is devalued in comparison to the increased "mass" of the Canadian dollar.

The General Theory Explains Gravity

Gravitational Waves Bending Light

There is a misconception about the general theory of relativity—namely, that few people in the world understand it because it's so difficult to comprehend. Though this statement might have had much truth to it when Einstein completed the general theory in 1916, it is no longer the case. Physics professors around the world teach the theory, scientific articles on general relativity number in the thousands, and numerous books for laymen are devoted to various aspects of the subject. However, the general theory of relativity does require familiarity with certain principles of mathematical physics and is more difficult to grasp conceptually than the special theory. Therefore, we will only briefly summarize it here.

Over the ten years following his 1905 paper on special relativity, Einstein expanded that theory into the general theory of relativity. He explained:

> The theory of relativity resembles a building consisting of two separate stories, the special theory and the general theory. The special theory, on which the general theory rests, applies to all physical phenomena with the exception of gravitation; the general theory provides the law of gravitation and its relations to the other forces of nature.

The general theory explains gravity beyond Newtonian physics. Newton explained that the mass of a body causes gravity and how to calculate the force. But Einstein explained *why* matter causes gravity. To understand the general theory, we have to go back to the frames of reference. Einstein asked what nature has to do with these frames of reference and their respective states of motion, and concluded that laws of motion must be entirely independent from the choice of the particular reference frame used to describe nature. This point—that the laws should be entirely independent of the choice—is called the general principle of relativity.

His search for unity and simplicity in scientific principles led Einstein to this theory, the same objective that first led him to the special theory and the reason he later pursued the unified field theory. As illustrated earlier in this chapter, application of the special theory is limited to frames of reference moving in a straight line toward or away from each other at a constant speed, that is, without acceleration. The general theory, on the other hand, provides a formula for the relationship of matter *throughout space moving in any direction, with or without acceleration.* It provides a unifying and simplifying principle that (1) modifies the principle of the constant velocity of light to address encounters with strong gravitational fields; and (2) accounts for the fact that all bodies (regardless of mass or composition) fall freely in a gravitational field with the same acceleration.

The main predictions derived from the general theory include these:

1. All electromagnetic radiation (including light) is deflected by gravitational force.
2. Mercury's orbit deviates from the orbit calculated by Newtonian physics.
3. A clock on the surface of a massive object will run slower than an identical clock in free space.
4. Gravitational *waves* exist, radiating at the speed of light, from large masses that are accelerating.

Numbers 1, 2 and 3 have been proven, while number 4 is the subject of intensive investigation by a large community of scientists at universities and research centers throughout the world. The end result of general relativity is a further extension of physics to account for gravity and accelerated frames of reference, and leads to the conclusion that space is "curved" and that not only are space and time linked, but that the complete explanation of the universe combines space, time, *and matter.*

Relativity Impacts Our Everyday Lives

An Aid to Navigation

As obtuse and esoteric as relativity is, it actually has some practical impacts on our lives. For example, Einstein's equation $E=mc^2$, which grew out of the special theory, opened the way for atomic research. In addition, special relativity is applied in nuclear medicine, the branch of medicine that uses radioactive materials for the diagnosis and treatment of disease. For instance, the element technetium is widely used in bone scans for cancer patients to determine whether the cancer has metastasized to the bone. Although technetium occurs in the Earth's crust in minute traces (from the spontaneous fission of uranium), it was first discovered in 1937 as a result of high-energy radiation of the element molybdenum in a nuclear particle accelerator. The construction of particle accelerators—and, thus, the creation of some isotopes used in nuclear medicine—involves an understanding of special relativity's mass/energy principles.

General relativity's ties to modern technology can be seen in the fast-growing use of the Global Positioning System (GPS), which is a precise satellite-based navigation and location system originally developed for U.S. military use. It was used in the Persian Gulf War in 1991 (Desert Storm) to locate land mines and lost personnel. GPS now has a wide variety of existing and potential non-military uses—including detection of tectonic plate movements and identifying the exact location of sea and land travelers. In addition, power companies use GPS to locate breaks in power lines, it is used to land aircraft, to trace the migratory paths of animals, and to locate survey benchmarks that became buried after many years. To maintain precision, GPS must incorporate adjustments to the clocks on the satellites (applying special relativity) and to the satellites' locations based on Einstein's general theory of relativity. With over 50 manufacturers of 350 products worldwide, consisting primarily of software utilizing GPS, relativity is becoming the basis for an entire new industry. It is likely that everyone will be significantly affected by GPS, directly or indirectly, within the next few years.

In 1919, three years after Einstein had completed his work on the special and general theories of relativity, he wrote:

> Let no one suppose . . . that the mighty work of Newton can really be superseded by this or any other theory. His great and lucid ideas will retain their unique significance for all time as the foundation of our whole modern conceptual structure in the sphere of natural philosophy.

Einstein modestly described the special theory of relativity as "simply a systematic development of the electrodynamics of Maxwell and Lorentz," which put the "finishing touch to [their] mighty intellectual edifice . . ." He also recognized the importance of the discoveries of Marie Curie and Ernest Rutherford, through which he became aware of the energy in matter. This, in turn, led him to $E=mc^2$. "We see atom disintegration," said Einstein, "where nature herself presents it, as in the case of radium."

Beginning with the moment he posed the innocent question at the age of 16, in 1895—*What would the world look like if I were riding on a beam of light?*—and culminating with the full development of the general theory of relativity in 1916, Einstein brought together the diverse concepts and discoveries of his predecessors, like discordant notes of musical instruments, until he formed a full symphony—a revolution in physics—that exploded with a mighty crescendo heard around the world. Thus, like Newton, Einstein stood on the shoulders of giants, and like Newton, his greatness exceeded those who came before him.

Though Einstein's fame rests on relativity, he wrote prolifically on numerous other scientific subjects in his early years. In the eleven years between his two major publications on relativity, he wrote sixty papers on such topics as photochemistry, quantum statistics of gas, thermodynamics, optics, molecular physics, electromagnetism, and the history and philosophy of science. Between 1915 and 1930 his work on quantum theory was highly influential. All together, his scientific publications number about 350.

On April 18, 1955, after a life that included two of the greatest contributions to 20th-century physics and that was intertwined with the rise and fall of Germany, the formation of Israel, and the creation of the atomic bomb, Albert Einstein died in his sleep at Princeton Hospital. Of the two passions that dominated his existence, he once said, "Politics are for the moment. An equation is for eternity."

Part Four

The Big Bang and the Formation of the Universe

In Focus

From Galileo's observations beginning in 1610 we gained but an inkling of the vastness of the heavens. The full story didn't unfold until the 20th century, when giant telescopes and other equipment allowed us to see the true size and nature of the galaxy that is Earth's home, then to see galaxies beyond it.

Our solar system is located in the outer reaches of a vast rotating galaxy—the Milky Way—and moves in relation to its center at a speed of 140 miles per second. One revolution of this group of three hundred billion stars takes 230 million years. The Milky Way Galaxy is but one of billions of galaxies, all of which are still reeling from a giant explosion called the Big Bang, which continues to show its effects today in the light spectra of distant galaxies and in the actual cosmic radiation that remains measurable throughout the universe. By measuring the speed of the galaxies' recession, we've been able to calculate that the entire universe was created about fifteen billion years ago—not only the formation and motion of the galaxies, but the creation of space-time itself.

Through a detailed understanding of the elementary particles that comprise atoms and by studying the cosmos, we've been able to determine the conditions that existed immediately after the explosion and the subsequent events that evolved from it. One result was the formation of the elements that make up Earth and its inhabitants. Thus, we can look back and see the creation of both the physical universe and life on this planet.

Two questions of immense proportions remain: Where did the "material" that exploded in the Big Bang come from? What is the ultimate fate of the universe? Part Four examines the discovery of the Big Bang and its implications, and explores possible answers to those two monumental questions.

Chapter Nine

The Cosmic Egg

"I have observed the nature and material of the Milky Way. With the aid of the telescope this has been scrutinized so directly and with such ocular certainty that all the disputes which have vexed philosophers through so many ages have been resolved. . . . The galaxy is, in fact, nothing but a congeries of innumerable stars grouped together in clusters. . . . And what is even more remarkable, the stars which have been called "nebulous" by every astronomer up to this time turn out to be groups of very small stars. . . . In the nebula called the Head of Orion, I have counted twenty-one stars. The nebula called Precipe . . . is a mass of more than forty starlets . . ."

GALILEO GALILEI
The Starry Messenger (1610)

"On the far side of the Big Bang is a mystery so profound that physicists lack the words even to think about it . . . The Universe consisted of . . . maybe only one particle that interacted with itself in that tiny, terrifying space. Detonating outward . . . every bit of matter was crushed with brutal force into every other bit, within a space smaller than an atomic nucleus."

ROBERT CREASE and CHARLES MANN
The Second Creation (1986)

Look up to the heavens. At night, about five thousand stars are visible to the naked eye. Until about four centuries ago, this seemed to be the extent of the universe—a finite number of unchanging, immovable stars spread evenly throughout a finite area. Except for the occasional nova like the one Tycho Brahe observed in 1572, the universe of stars surrounding our solar system appeared much as Aristotle had described it more than 2,000 years ago. But, as we saw in Chapter 2, in 1609 Galileo improved the telescope and began to study the band of constellations that included Orion, Perseus, Cassiopeia, Cygnus, Aquila, Sagittarius, Centaurus, and Carina. This band was likened to a river of milk by the ancient Greeks because of its light, cloudy appearance, and was later called the Milky Way. Galileo saw that the Milky Way was not comprised of "clouds" at all. It was a vast collection of stars. Thus began the Renaissance revolution beyond Copernican theory and the solar system, as it extended to the starry heavens. Centuries later, this revolution revealed not only what exists in the heavens, but also unveiled an extraordinary and intricate pattern and story that no one could have ever imagined—a story that tells how the heavens came to be the Sun, Earth, planets, stars, and the very matter from which life was created.

Astronomers Discover Mankind's Galaxy and Location in It

In and Out of Spiral Arms

In 1781, the self-taught British astronomer Sir William Frederick Herschel (1738–1822) discovered the planet Uranus, the first planet to be discovered since prehistoric times. Then in 1785, after carefully cataloging the positions of thousands of stars, Herschel was the first to suggest that our solar system is actually part of a larger system of at least several million stars, and that it has the shape of a "thin disk of nearly infinite extension." After twenty more years of systematic observation with the most powerful telescopes in existence, which he constructed himself, Herschel cataloged 2,500 star clusters in the Milky Way Galaxy.

The quality of telescopes continued to improve dramatically in the 1800's, enabling other astronomers to see millions of stars never before observed, allowing for improvements in measuring the distance of stars from Earth, and leading to the discovery of the eighth planet, Neptune, in 1845. In the early decades of the 20th century, based on work performed by Harlow Shapley (1885–1972) and Robert J. Trumpler (1886–1956), we gained our modern understanding of the Milky Way Galaxy. Their work demonstrated that our solar system is located in a spiral arm near the edge of this giant rotating system of stars, which measures about 100,000 light-years across (see Figure 9-1). The center of the galaxy is located approximately 27,000 light-years away from us. *That is, traveling at 186,000 miles per second (the speed of light, traversing about 5.88 trillion miles in a year), it would take 27,000 years to get from Earth to the center of the Milky Way Galaxy.*

Figure 9-1

MILKY WAY GALAXY

Edge-on view   **Face-on view**

As astronomers continued to map the heavens, they began to realize the incredible vastness in which we live and that William Herschel's estimate of at least several million stars in the Milky Way was a gross underestimate. We came to see that in this galaxy alone there are actually about 300 billion stars. Of them, we are near only one. Our Sun is 93 million miles from Earth. It is an average-size star, but accounts for 99 percent of the total mass of the solar system. This system of planets and other orbiting material (comets, asteroids, dust, gas, and other space debris) is located in one of the spiral arms and orbits the center of this gigantic mass of stars, with the 300 billion stars. The star closest to our solar system is called Proxima Centauri, and is located about four light-years away (23 trillion miles), in the same spiral arm. The galaxy also contains enormous hydrogen and dust clouds that astronomers are still in the process of detecting and measuring. This material extends so far beyond the visible stars that if it were included as part of the Milky Way, the galaxy would be about 500,000 light-years across (rather than 100,000).

Our solar system and this area of the spiral arm move at a speed of 140 miles a second—that is, about 500,000 miles per hour in relation to the center of the galaxy. One revolution takes about 230 million years. The Sun has made twenty revolutions around the center of this immense galaxy. If we were to view this rotation from a distance and speed it up, we'd see the galaxy pass through various stages in this ponderous motion and evolution. Just as the planets revolve about the Sun with the more distant planets moving slower than Earth, Venus, Mars and Mercury, the outer areas of the Milky Way rotate progressively more slowly.

Astronomers Discover a Hundred Billion More Galaxies

The Starry Messenger Revisited

Is this vast revolving cluster of 300 billion stars the entire universe? Does this collection of burning hydrogen and helium spheres along with Earth and some planets and other matter void of life comprise all that exists? Infinite space beyond?

In A.D. 964, the Arab astronomer al-Sufi wrote *Book of the Fixed Stars*, and was the first to note a small object that looked like a fuzzy star. In the year 1612, the German astronomer Simon Marius (1573–1624), using the newly invented telescope, rediscovered this faint area far out in space, saying it looked like "the light of a candle seen through a horn." It came to be called the Andromeda Nebula, and was believed to be a luminous cloud of gases and dust in the Milky Way Galaxy. Despite subsequent improvements to the telescope, individual stars in the Andromeda Nebula were not visible until 1885, when a nova appeared in Andromeda for several days. The nova indicated that Andromeda was more than just dust and gases—at first, it was believed to be a *cluster of stars* on the distant fringes of the Milky Way.

American astronomer Edwin Powell Hubble (1889–1953), the founder of extragalactic astronomy and namesake of the Hubble Space Telescope launched in 1989, pioneered the research on Andromeda. After earning degrees in math and astronomy in 1910 at the University of Chicago, he studied law as a Rhodes Scholar at Oxford and began practicing law in Kentucky in 1913. Becoming bored with the legal profession, Hubble abruptly changed directions once again, and earned a Ph.D. in astronomy at the University of Chicago in 1917. He later went to work at Mount Wilson Observatory in California, where the new 100-inch telescope finally made it possible to see the vast number of individual stars in Andromeda.

Hubble's work at Mount Wilson showed that Andromeda was not just gases and dust and a few novas—it was made up of stars, billions of them. Even more amazing, it wasn't in the Milky Way at all. It was far, far outside this galaxy. By 1923 Hubble had determined that Andromeda is another galaxy in its own right, an independent spiral galaxy about 200,000 light-years in diameter and about *two million light-years* from the Milky Way. Andromeda was the first galaxy proved to be beyond the Milky Way. The *nearest* large galaxy, Andromeda is the *farthest* object that can be seen with the naked eye. Figure 9-2 is a photograph of Andromeda and its two companion galaxies (called NGC 205 and M32) taken by the 48-inch telescope at Mount Palomar Observatory northeast of San Diego, California.

Hubble further determined that thousands of other "nebulae" are also galaxies, much like Galileo had written of the Milky Way nebulae more than 300 years earlier. Thus, the map of the universe, and mankind's concept of where we fit in that universe, had to be redrawn once more. *Since 1925, astronomers have found that there are about 100 billion separate galaxies in the observable universe, averaging 100 billion stars each.* The *farthest* known galaxy from Earth, 4C41.17, is about 15 billion light-years away. The *largest* discov-

Figure 9-2

THE ANDROMEDA GALAXY

ered so far, called "Abell 2029," is about 60 times the size of the Milky Way, with over 100 trillion stars and a diameter of six million light-years. Like the Milky Way, most of these groups of stars are shaped like whirlpools of light, disclike structures with spiral arms. About 90 percent of all the galaxies in the universe can be classified into one of the three broad categories shown in Figure 9-3: elliptical, spiral, or barred spiral.

There are single galaxies, galaxy pairs, and large galaxies with smaller satellite galaxies. Most galaxies are part of "clusters" of galaxies, averaging about 100 galaxies per cluster. In the constellation Coma Berenices there is a cluster of about 10,000 individual galaxies. The Milky Way is part of a cluster of 20 galaxies. The clusters are distributed throughout space in groups that form a generally uniform or homogeneous pattern, with an average of about 10 million light-years between each group.

Figure 9-3

GALAXY SHAPES

Spiral Galaxy Messier 101
Located in Ursa Major Constellation (the "Big Dipper")

Elliptical   Spiral   Barred Spiral
Hubble Classification Scheme of Galaxies

Hubble Calculates the Recession Speed of the Galaxies As the Universe Rushes Apart

Sirius and the Red Shift

In the late 1800's, astronomers and physicists began to develop a method to determine stars' and other heavenly bodies' movement toward or away from Earth by measuring the light spectra of such bodies as we perceive their light here on Earth. English astronomer William Huggins (1824–1910) and French physicist Armand-Hippolyte-Louis Fizeau (1819–1896) discovered that certain spectral lines of some of the brighter stars are "shifted" slightly to the red from their normal position in the light spectrum. They correctly concluded that this was similar to the Doppler effect (named after the Austrian physicist Christian Doppler) with sound waves. The Doppler effect explains the phenomenon of sound waves being "compressed" together when the source of the sound (such as a train whistle or ambulance siren) is moving toward the listener—causing an increasingly higher pitch—and the waves being "stretched" apart when the source is moving away from the listener, causing an increasingly lower pitch. Although light is made of photons and sound results from air vibrating, the two are similar in the sense that both come in wavelengths that can be measured. *A change in the light wave toward the red (lower frequency or longer wavelength) results from the star moving away from the observer on Earth.* In 1869 Huggins determined that Sirius (the brightest star in the northern sky) is moving away from the Earth at about twenty miles a second.

Edwin Hubble took this information about the shift in light spectra and made his second remarkable discovery, one that would turn out to be the greatest discovery of 20th-century astronomy. In 1927, combining the previous research on light spectra with the powerful new telescopes at Mount Wilson, Hubble discovered that the red shift of the receding galaxies increases in proportion to the distance from us. In other words, the universe is expanding, with the farthest stars moving faster. The rate of expansion is represented by his calculation that is now called Hubble's constant: *Recession speed equals distance divided by a time factor of about 15 billion years. Every galaxy cluster (but not necessarily every galaxy) is rushing away from every other cluster of galaxies, precipitously, at varying velocities in accordance with Hubble's constant, with the galaxies at greater distance receding at a faster rate.* Today's calculation of Hubble's constant shows that the galaxies are expanding at a velocity of about 10 to 20 miles per second for every million light-years in distance from Earth. For example, the Ursa Major II cluster of galaxies is receding at 25,930 miles a second, 14 percent of the speed of light. In the universe's outermost regions, galaxies have been clocked at 84 percent of light speed.

The expansion discovered by Hubble is the space *between* the clusters of galaxies, not *within* galaxy clusters. Each galaxy and galaxy cluster is bound together by gravity and is not expanding. So, for example, the Andromeda Galaxy and the Milky Way, part of one cluster in which the galaxies are gravitationally attracted, are actually moving *toward* each other and will collide or coalesce in hundreds of millions of years. Throughout the universe in general, however, a rapid expansion is taking place. As we discuss in the next several pages, there is an inescapable connection between that expansion and the very creation of the universe.

By Calculating the Universe's Expansion in Reverse, We Are Led to the Cosmic Egg, the Big Bang, and Creation

Zero-Time at a Hundred Trillion Degrees

Until the last century, theories about the beginning of time tied the creation of humans and all living things to the creation of the universe. Irish Bishop James Ussher (1581–1656) wrote a chronology of the Old Testament, in which he added up all the generations of men and women mentioned in the Bible since Adam and Eve, and pegged Creation at 4004 B.C., *at 2:30 p.m., Sunday, October 23!* One person is said to have asked the bishop, "And pray, Holy Father, what was God doing *before* he created the universe?" To which Ussher replied impatiently, "Creating hell for those who ask such questions!"

An understanding of the receding galaxies through 20th-century astrophysics has allowed us to look back in time to the actual creation of the universe in an event that Bishop Ussher could never have dreamed possible. If we imagine the expansion of the universe in reverse and mathematically calculate its motion in reverse, all the galaxies would simultaneously meet at a point. This is called "zero-time" and considered the beginning of the universe. Although there is still some question about how long ago that was, most experts agree that zero-time was about 15 billion years ago.

"Once upon a time there was no time," wrote John Barrow in *The Origin of the Universe*. What happened at zero-time and thereafter to bring us to where we are today? In 1927, after becoming aware of Hubble's discovery of the expanding universe, the Belgian astronomer Georges Edward Lemaitre (1894–1966) proposed the theory that is now generally accepted by astronomers and other experts. At zero-time, he said, all matter was in one tiny mass that he called the "super-atom" or the "cosmic egg." Nothing else existed—no galaxies, planets, stars or life. Indeed, the elements didn't exist because the cosmic egg was composed of energy not yet even in the form of nuclei or atoms. Without matter, there was no time. The cosmic egg was subject to its own gravitational pull and contracted and grew closer and closer together, creating higher and higher temperatures as it was compressed into a smaller and smaller volume. At that moment, the entire universe was a "kernel" of energy. At some point, with intensely high temperature at the smallest volume possible, a great explosion occurred, and this single kernel of energy became everything there is. Lemaitre said that the recession of the galaxies is the visible evidence of this explosion. His theory also explained the recession of the galaxies within the framework of Einstein's general theory of relativity.

The explosion is the earliest event about which we have any record. The theory was further refined and modified by George Gamow (the Russian-born physicist responsible for the "liquid droplet" model of the atomic nucleus, discussed in Chapter 6) and published in a 1948 paper titled "The Origin of Chemical Elements," in which he postulated that the elements were formed by atomic nuclei being built up by the successive capture of neutrons, and in which he first coined the term "big bang." The "Big Bang theory" was then made popular in 1950 by the British astrophysicist Fred Hoyle in a series of BBC radio talks entitled "The Nature of the Universe." It is so widely accepted that astronomers now call it the "Standard Model."

We now have a surprising amount of reliable circumstantial evidence about the Big Bang. Steven Weinberg (1933–), then a physics professor at Harvard University, received the 1979 Nobel Prize for

Physics for his work in this field. He described the results of that research in his book *The First Three Minutes:*

> In the beginning there was an explosion . . . At about one hundredth of a second [later] . . . the temperature of the universe was about one hundred trillion degrees Centigrade. This is much hotter than in the center of even the hottest star, so hot, in fact, that none of the components of ordinary matter, molecules, or atoms, or even the nuclei of atoms, could have held together. Instead, the matter rushing apart in this explosion consisted of various types of the . . . elementary particles . . .
>
> As the explosion continued the temperature dropped . . . reaching one trillion degrees at the end of the first three minutes. It was then cool enough for the protons and neutrons to begin to form into complex nuclei, starting with the nucleus of heavy hydrogen . . . which consists of one proton and one neutron. The density was still high enough . . . so that these light nuclei were able to rapidly assemble themselves into the most stable light nucleus, that of helium, consisting of two protons and two neutrons. . . . At the end of the first three minutes the contents of the universe were mostly in the form of light, neutrinos and anti-neutrinos . . . This matter continued to rush apart, becoming steadily cooler and less dense. Much later, after a few hundred thousand years, it would become cool enough for electrons to join with nuclei to form atoms of hydrogen and helium.

Although the Big Bang occurred about 15 billion years ago, it took several billion more years for the galaxies to acquire their present configuration in the universe, including the Milky Way and our own solar system within it. The hydrogen and helium rotated increasingly faster, flattening along the axis of rotation. Like the Milky Way, most of it formed into disclike structures with spiral arms—great wheels of matter and light—as well as the other galaxy configurations shown in Figure 9-3.

Here, in the outer reaches of the Milky Way Galaxy, about four and a half billion years ago a cloud of hydrogen and helium began to condense and fall inward upon itself because of its own gravitational pull. The matter from which it was formed could have come from the explosion of a supernova, rather than forming at the Big

Bang. In either case, this matter first formed a flat, spinning disc, like the Milky Way itself. But as more and more material collected, the center of the disc grew thicker, and the greater gravitational pull from the enlarging object caused more space debris to be attracted until the object formed into the most compact geometric shape—the sphere—and became so massive that its interior was under incredible physical pressure. That pressure caused the new sphere to heat to temperatures of up to 20 million degrees, which caused the hydrogen atoms to fuse into helium in a continuous and massive series of nuclear reactions, marking the birth of our Sun. This process was repeated in the birth of every one of the hundreds of trillions of other stars in the universe. As discussed in Part Two and Part Three, when this fusion takes place we see—and we exist as the result of—the transformation of mass into light and heat.

Experts differ on whether the universe will continue to expand indefinitely. All the galactic clusters that are rushing apart appear to be slowing down. Nevertheless, the explosion might have catapulted this material with sufficient energy to escape from the gravitational pull of the rest of the material. If it did, the universe will expand forever. On the other hand, if the material is not traveling with such energy, all of the universe will fall back on itself someday and perhaps create another big bang and begin the process all over again. Even if this were to occur, it would be tens of billions of years before the universe meets that ultimate fate.

Heavier Elements Coalesce to Form One Percent of the Universe, Including Earth

Pluto and the Oort Cloud

As Steven Weinberg explained, we now understand the process that led to the formation of the elements during and after the Big Bang. The cosmic egg predominantly formed into hydrogen atoms (which contain only one proton and one neutron), followed by the next simplest atom, helium (with only two protons and two neutrons). Today, these two elements comprise 99 percent of the entire universe (74 percent hydrogen and 25 percent helium). Over the succeeding hundreds of millions of years, the heavier elements that comprise the remaining one percent of the matter in the universe formed as an offshoot of early stars' fusion reactions and as a result of catastrophic stellar explosions. In a relative sense, there has been a very, very small nuclear change, or change in the elemental makeup of the universe, from the time immediately after the Big Bang until now. Trillions times trillions times trillions of interactions of hydrogen atoms, helium atoms, and other elementary particles occurred to form elements other than hydrogen or helium—*yet, those 90 other naturally occurring chemical elements comprise less than one percent of the entire universe.*

Focusing for a moment on *living matter on Earth,* it is also remarkable that 99 percent of the material making up all living organisms is made of just *four* types of atoms: hydrogen, carbon, nitrogen and oxygen. It's likely that every atom in your body was once part of a star. To form carbon, for example, three helium nuclei typically collide with each other simultaneously in a millionth of a millionth of a second. With a couple of relatively minor exceptions to this process, the carbon atoms in every organic molecule in every cell in every living creature on Earth were formed by such triple collisions—part of the spectacular transformation from elementary particles to living beings that can now comprehend from where we came.

As mentioned earlier, matter gravitationally "pulls" itself into spheres because that geometric figure possesses the smallest surface (i.e., the most "efficient" configuration) for a given total volume of matter. Thus, around the same time as the Sun was forming, dust and gas that included heavier elements were floating through our galaxy and began to coalesce into larger masses attracted to the Sun by its enormous gravitational pull. As the Sun condensed, so did these other masses of matter, eventually forming more solid bodies, which collided with each other over a period of millions of years and combined into larger solid spheres. This coalescing of matter and succession of collisions formed planet Earth, the other planets, and the numerous moons of planets in our solar system.

Initially, the Earth was extremely hot and had no atmosphere. An early atmosphere was then formed by the emission of hydrogen sulfide and other gases from the molten materials. Eleven billion years after the Big Bang, the Earth's primordial soup gave rise to the first organic molecules. Scientists continue to expand their knowledge of the conditions that existed in the earliest years of Earth. This research includes gathering information from other planets. In 1995, after a decade of planning and traveling to its destination, a 746-pound satellite probe from the $1.35 billion Galileo spacecraft entered Jupiter's atmosphere and radioed back a wealth of data for fifty-seven minutes before it plunged into the planet and was crushed and melted by Jupiter's immense gravitational force and heat. This measurement of oxygen, hydrogen, and other gases and temperatures by the probe's ultraviolet spectrometer gave us important information about *Earth's* early moments of formation billions of years ago, because Jupiter is so massive and its gravity so great that its atmosphere is virtually the same as when the entire solar system was formed four and a half billion years ago. Those same gases formed the basis for Earth's origin, but were gradually altered by their escape to space, the bombardment of meteors and comets, chemical reactions, and volcanic activity. In contrast, although Jupiter has undergone its own pressure and chemical changes, it remains a relatively pristine example of the original primordial cloud and provides valuable clues to Earth's evolution through comparison of the two planets.

Enormous amounts of debris in our solar system never coalesced into planets or moons—these are the asteroids, comets, and other dust and material, which are particularly common around the outer planets. Asteroids are made of rock and metal and come in all sizes, from a few ounces to millions of tons, and the vast majority (about 7,000) orbit the Sun in an asteroid belt that exists between Mars and Jupiter. The largest known asteroid, Ceres, is 581 miles in diameter. Pieces of asteroids result from collisions between them, and a number of those pieces strike the Earth every year in the form of meteorites. Comets, on the other hand, are primarily composed of water, ice, hydrogen, carbon, oxygen, and nitrogen, and number in the trillions in our solar system. This material forms a "tail" of vapor, gas, and dust—hence, the term comet, from a Greek word meaning "long-haired." Nearly all comets revolve around the

Sun in a plane similar to the plane of the major planets, and often cross planets' paths, as seen with Halley's Comet. Some comets' orbits are short (within the area between the Earth and the Sun), while others have tremendously elongated paths that extend far beyond Pluto's orbit. Most astronomers subscribe to the theory that comets originate in a vast cloud of material called the Oort Cloud, named after the Dutch astronomer Jan Hendrik Oort (1900–1992), who first proposed the theory of the existence of this material in that region and how it results in comets.

Astronomers Discover Other Solar Systems in the Milky Way Galaxy

Life on Mars

Beginning in the early 1980's, improved methods to search for other solar systems have led to the discovery that planets orbit other stars in the Milky Way. The first such discovery took place in 1992, when astronomers found a star with two planets 1,300 light-years from Earth. In October 1995, astronomers detected a planet about half the size of Jupiter orbiting a star called "51 Pegasi" in the constellation Pegasus. In January 1996, two California astronomers detected two planets larger than Jupiter accompanying the stars 70 Virginis, in the constellation Virgo, and 47 Ursae Majoris in Ursa Major (the Big Dipper), about 35 light-years away. Trillions of miles from Earth, these planets are too distant to be seen with telescopes, but sensitive instruments aboard satellites and powerful Earth-based devices can monitor stars to see if they wobble, indicating gravitational pull by a planet. Some of these instruments measure the pattern of infrared light in the thin, gaseous, pancake-like disc around the star to detect the ring-shaped gap that any revolving planet would sweep out.

The temperatures of the planets in Virgo and Ursa Major are between 100°F below zero and 185°F above zero—a range conducive to the chemical processes that could allow for life to be formed and sustained. Researchers have also found ordinary vinegar in a stellar cloud 25,000 light-years from Earth, by analyzing the unique spectrographic "signature" of its wavelengths of light and radio signals from space. Vinegar consists of an organic molecule (acetic acid) that might have played a role in the formation of life on Earth. The same technology has been used to identify all four elements critical to the formation of life on Earth—hydrogen, carbon, nitrogen and oxygen—in massive clouds of dust and gas throughout the universe.

The speculation about life elsewhere in the universe took a startling twist in August 1996 when NASA announced the discovery of fossilized organic molecules and possible cells in a meteorite from Mars. The rock had been catapulted from Mars about 15 million years ago when a huge asteroid collided with the planet. It landed on Earth about 13,000 years ago, and contains polycyclic aromatic hydrocarbons (PAH's), which some biologists say is unequivocal evidence of the presence of water in the early history of Mars (about three billion years ago) and the *presence of biological organisms* that produced the PAH's at that time—that is, about the same time that life was first forming on Earth.

In light of the existence of trillions of stars, the recent discoveries of planets around some of those stars, and the possibility of past life on Mars, it appears that our solar system and perhaps even the existence of life on Earth are not unique. It is beginning to seem highly unlikely that one ordinary star, our Sun, is the only star accompanied by a planet that sustains life. If life or past life is discovered on Mars or elsewhere, humanity's sense of itself will be shaken once again. Amid the developing technology, astronomers are predicting the dawn of a new era in astronomy, the second coming of Columbus wherein we will find new worlds. Within four months after the discovery of the Mars meteor, three spacecraft were launched separately by the United States and Russia to study the planet's atmosphere, map its surface and identify areas that might have hosted life. Future missions will bring back Martian rock and soil. Perhaps there are millions of life-sustaining bodies out there. In those planetary systems that we cannot see, but which exist, intelligent life might have evolved. Not only are we dealing with immense numbers of stars and potential solar systems, we must also consider this in the context of the fifteen billion years during which the universe has existed. Living systems, millions of them, could have evolved on other planets and could be scattered throughout the universe, some overlapping in their evolution and stages of development, with the birth of others coming millions of years after the death of ones preceding. After all, there was a time when Earth itself was totally inhospitable to any form of life. The environment evolved, and the planet gave birth. Our time as a product of this changing planet is but a second in infinity. As we will see in Part Six (on the cell), with the proper conditions, life could form anywhere in the universe. Life on another planet could have its time, too. It could be similar to life on Earth or it could consist of entirely different forms suited to that planet alone, passing through its multibillion-year cycle, with the planet then becoming inhospitable, desolate, dead once more. There could be life on a billion planets, and no living, thinking being from one planet would know first-hand about another.

As suggested by the question asked of Bishop Ussher, without cause there is no effect. Despite Aristotle's belief that the universe always existed, there is no physical phenomenon in our knowledge and experience of the universe that proves that energy or matter can come about spontaneously. Based on Einstein's equation $E=mc^2$, we are reminded that mass and energy are interchangeable. Neither energy nor matter can be created from *nothing*. Since it is clear that the universe is expanding, with all galaxy clusters moving away from all others, and since it is also clear that such movement is slowing down at a finite rate, the universe could not have existed forever in its current state. That is, the universe can't be infinitely old because it would have died already. Therefore, the universe must have come into existence a finite time ago. Thus, the question must be: Where did the cosmic egg come from? What was the "First Cause"? How did Lemaitre's super atom—that primordial egg of incredible density and ultimate temperature—come to be in the first place, and why did it explode?

The implications of these questions are immense. Perhaps someday a scientific theory will provide the possible answers. Some scientists have suggested that the cosmic egg was in existence for eternity before the Big Bang. Other people postulate that the cause of the cosmic egg or the cosmic egg itself was a supreme being or other supernatural force. Some cosmologists have developed theories that don't rely on the common understanding of cause and effect, because the same physical laws don't apply when there is no time, no space, and no matter. But assuming the usual meaning of cause and effect, we could easily conclude that such a supernatural force *must* have created the "stuff" that made up the cosmic egg. There is ample room in astrophysics for a supreme being at creation. But then, is there any other compelling or inescapable conclusion or belief that results from this fact or assumption? That is, even if everyone agreed that a supernatural force was the cause of the cosmic egg, are any further empirical conclusions derived from that "fact"?

The various creation myths of ancient times attempted to provide an explanation for our existence, to remove the unknown and provide psychological security and stability to society. Even though the Big Bang and the approximate time it occurred have been proven well beyond debate, millions of people continue to subscribe to the biblical precepts of spontaneous creation of the universe by God about 5,800 years ago. The evidence supporting the Big Bang is overwhelming. In light of that evidence, hope and belief in the eternal "truths" about creation that were conjured up centuries ago is far more dangerous to the progress and well-being of humankind and far less rational than it was when people allowed Aristotle's and Ptolemy's picture of the universe to be carried with such force into the Renaissance.

Beginning immediately after the first instant of creation, the universe evolved in a predictable fashion. Thus, even if one assumes that a supernatural being or force was responsible for creation of the universe at the time of the Big Bang, divine intervention is not needed for the long list of questions that we began to ask and have now answered in our more recent history. In *A Brief History of Time*, Stephen Hawking puts it this way:

The whole history of science has been the gradual realization that events do not happen in an arbitrary manner, but they reflect a certain underlying order, which may or may not be divinely inspired . . . These laws may have originally been decreed by God, but it appears that he has since left the universe to evolve according to them and does not now intervene in it.

Unresolved mysteries persist. But they are contained within a resolved framework. From this small planet in the outer reaches of an unremarkable galaxy, men and women have examined the vast space that surrounds them and unraveled a most magnificent description of creation.

Chapter Ten

The Echo of Creation

"The Earth is a place. It is by no means the only place. It is not even a typical place. No planet or star or galaxy can be typical, because the Cosmos is mostly empty. The only typical place is within the vast, cold, universal vacuum, the everlasting night of intergalactic space, a place so strange and desolate that, by comparison, planets and stars and galaxies seem achingly rare and lovely."

CARL SAGAN
Cosmos (1980)

The first chapters of the Big Bang theory were written by astronomers whose great telescopes explored the sky during the first few decades of this century. Then a startling and unexpected new chapter was written by those in a new field of astronomy—the radio astronomers—when they discovered complex clues scattered throughout the universe fifteen billion years ago. Not only can we *see* the evidence of the Big Bang through the galaxies' recession, we can actually measure the remnants of the "heat" left over from that explosion.

As we learned in the previous chapter, the early universe was incredibly hot. The exceedingly high temperature was significant not only because of its effect on atomic particles, but also because the entire universe was bathed in electromagnetic radiation, which *cooled* along with the matter. To understand how the detection of this radiation led to the confirmation of Hubble's observations and Lemaitre's Big Bang Theory, we must first define electromagnetic radiation.

Maxwell Predicts the Existence of Invisible Waves

Volts, Amps, and Coulombs

As suggested by the term itself, electromagnetism refers to one force or phenomenon that is a combination of electricity and magnetism. Following Newton's example of applying the scientific method to the branch of physics called mechanics, 18th-century and 19th-century researchers like the French physicists Charles Coulomb (1736–1806) and André-Marie Ampère (1775–1836), the Italian physicist Alessandro Volta (1745–1827), the German mathematician Karl F. Gauss (1777–1855), and the English physicist and chemist Michael Faraday (1791–1867) performed hundreds of experiments with electricity and magnetism in an effort to understand these phenomena. Their work led to an understanding about the nature and characteristics of electricity and magnetism, which was elaborated upon and brought together in 1861 in a set of four statements or laws by the Scottish physicist James Clerk Maxwell (1831–1879). These laws and their supporting equations completely define the relationship between electric and magnetic fields and play the same role in electromagnetism that Newton's laws do in gravity and motion. They can be summarized in this fashion:

1. An electric charge produces an electric field.
2. A magnetic field exists between the poles of a magnet.
3. Electric fields are produced by changing magnetic fields.
4. Magnetic fields are produced both by changing electric fields and by electric currents.

The first two principles explain *static* electric and magnetic fields, that is, fields without any currents or changing currents. Maxwell's most significant contribution was to the fourth principle. He recognized that magnetic fields are not only produced by electric currents, but are also produced by changing electric fields. After he set forth and elaborated on the four equations, he realized that laws three and four mean that propagating electric fields and magnetic fields cannot be separated because each produces the other. From this realization, he predicted the existence of *largely invisible energy waves*, which we now call electromagnetic radiation. That is, based fundamentally on the third and fourth equations, Maxwell predicted the existence of oscillating electromagnetic "fields" moving through space like waves or ripples in a pond radiating from their source.

Electromagnetic waves possess three characteristics: wavelength, speed, and frequency. As seen in Figure 10-1, each wave consists of a series of crests and troughs. Wavelength is the distance between adjacent crests. Speed is a measurement of the movement of the crests. Frequency is a quantity that indicates how many crests pass a given point in one second (usually measured in "hertz," after the German physicist Heinrich Hertz, discussed below).

Figure 10-1

THE ELECTROMAGNETIC SPECTRUM

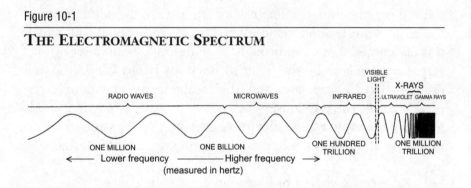

Light is the narrow band of electromagnetic radiation that is visible to humans. Maxwell calculated that the speed of all electromagnetic waves (including light) is 186,000 miles per second. This led him to the postulation that these waves exist at all frequencies and wavelengths. The wavelength depends on the frequency or rate that the field is oscillating—the higher the frequency, the shorter the wavelength. Thus, the waves of electromagnetic radiation cover the entire spectrum of radiation, with only a small portion (with wavelengths between 16 and 32 millionths of an inch) visible to humans in the form of light. As shown in Figure 10-1, the electromagnetic spectrum includes everything from radio waves (with wavelengths of a few yards to thousands of miles) to microwaves (one tenth of an inch to a foot) to X-rays and gamma rays, the most energetic electromagnetic radiation that we can measure, which is produced in stars and as a result of the decay of some radioactive elements.

As we saw in Chapter 8, Einstein later relied on Maxwell's work in arriving at the special theory of relativity. He described Maxwell's theories as "the most profound and the most fruitful that physics has experienced since the time of Newton." Also, the knowledge of spectral lines enabled Edwin Hubble to detect the red shift of the galaxies moving away from one another, as discussed in Chapter 9.

Though Maxwell's equations led him to the prediction of the existence of electromagnetic radiation, it took another twenty-five years for that theory to be generally accepted—primarily because no such waves had ever been proven to exist or had ever been created in the laboratory until the German physicist Heinrich Hertz (1857–1894) did so in 1887, eight years after Maxwell's death. Hertz was the first person ever to broadcast and receive radio waves through ingenious experiments using the first primitive versions of transmitters, antennas, and detectors of radio-frequency electromagnetic radiation. This signaled the birth of radio communication, and, of course, all of us now put electromagnetic waves to use in some practical way on a daily basis through radio, TV, computers, cellular phones, and many other devices. As stated earlier, Hertz became the namesake for the unit of measurement for wave frequency.

Penzias and Wilson Accidentally Discover the Cosmic Background Radiation

Missing an Antenna

We must now pick up the story begun by Faraday, Maxwell, and Hertz fifty years later. As radio technology developed, scientists found that the Sun's rays caused unwanted natural radio interference. In the late 1920's, Bell Telephone Laboratories wanted to set up a radio-telephone communication system across the Atlantic to supplement its cable system, and it needed to eliminate the interference.

So Bell launched a research program to better understand the manner in which the Sun causes the interference. The company constructed an array of antennas, receivers, and recording devices, which led to the discovery that radio emissions were not only coming from the Sun, but also from celestial bodies throughout the universe. The announcement of this discovery on April 27, 1933, gave birth to the modern science of radio astronomy. This represented a major advance in astronomy, because millions of celestial bodies emit radio waves that can now be detected by radio telescopes even when optical telescopes cannot see such bodies. Astronomy is the oldest science, begun in China, Egypt, and Greece thousands of years ago. For the first time in its long history, astronomy was no longer solely dependent on optical observation. We will now see that the advent of radio astronomy led to a monumental discovery that even went beyond the detection of previously undetectable stars and galaxies.

Through the 1930's and 1940's, astrophysicists developed an understanding of the precise conditions that existed after the Big Bang. They realized that the enormously hot early universe gave rise to a thermal electromagnetic radiation field, including X-rays and gamma rays, that is, radiation at the shorter wavelengths. As the universe cooled, the *average* temperature throughout it corresponded to longer and longer wavelengths on the spectrum. In 1948, physicists Ralph Alpher (1921–) and Robert Herman (1914–1997), working with the Russian-American physicist George Gamow (1904–1968), predicted that if the universe began in such a hot dense state, *there should still be radiation or energy with an average temperature of five degrees above absolute zero (on the Kelvin scale) remaining from the Big Bang and spread thinly throughout the universe.* They made this prediction in the same paper mentioned in Chapter 9, "The Origin of Chemical Elements." Because there was no equipment to detect such residual radiation, their prediction went unnoticed for nearly two decades.

In 1965 German-American astrophysicist Arno Penzias (1933–) and American radio astronomer Robert W. Wilson (1936–), working as radio engineers at the Bell Telephone Laboratories in Holmdel, New Jersey, were calibrating a sensitive radio antenna that was being used to communicate with the first Telstar communications satellite. They were attempting to find and eliminate the source of unwanted microwave "noise" that seemed to be registering with uniform intensity from all directions in space. The existence of these microwave signals (electromagnetic radiation with wavelengths between those of radio waves and infrared light) was very similar to the problem that Bell Laboratories faced in the 1920's, which led to the science of radio astronomy. Although this new Bell antenna was designed to track satellites, it continued to pick up this pervasive radiation regardless of the Earth's rotation. This uniformity indicated to Penzias and Wilson that the radiation could not be coming from a particular celestial body or galaxy.

In the meantime, at nearby Princeton University, P.J.E. Peebles (1935–), a theoretical physicist, was working with a group on the radiation issue first addressed by Gamow, Alpher, and Herman in 1948. Peebles and the Princeton group had recalculated the Gamow-Alpher-Herman 1948 postulation entirely independently. Since they weren't aware of any equipment capable of detecting cosmic background radiation, they began to design an antenna intended to do so.

Back at the Bell facilities, Penzias and Wilson remained at a loss to explain the source of the static they found with the Telstar antenna. Coincidentally, in a telephone conversation with the well-known astronomer Bernard Burke (1928–) of MIT in Boston concerning an unrelated matter, Penzias happened to mention the problem of the unexplained microwave signals. Being aware of the Princeton work, Burke realized that Penzias and Wilson must have found the cosmic background radiation predicted by Peebles at Princeton, and he introduced the two groups to one another. Penzias and Wilson and the Princeton physicists began working together, then simultaneously published papers in 1965 in the *Astrophysical Journal* that detailed their respective discoveries. Penzias and Wilson shared the Nobel Prize for Physics in 1978.

This marked a new era in research on the Big Bang and culminated in the information obtained by NASA's Cosmic Background Explorer (COBE) satellite in 1992. The equipment aboard COBE not only achieved the greatest accuracy to date on the temperature of microwave background radiation, it also detected the minute variations (one part in 100,000) in such radiation. This information has allowed cosmologists to recreate the approximate distribution of matter when the universe was only a million years old, which, in turn, has enabled them to figure out the events that led to the formation of the galaxies through basic gravitational forces, as explained in Chapter 9.

Not only were hydrogen and helium forged in the inferno of the Big Bang, an invisible, steady, and pervasive field of microwave radiation—a fundamental form of energy moving through the cosmos at the speed of light—was caused by the incredible heat of that explosion. Georges Lemaitre, the person who first proposed such an event in 1927 and lived to see it detected by radio astronomy, called this radiation "the vanished brilliance of the origin of the worlds." Theories and evidence accumulated in discrete and diverse quarters in the minds and laboratories of Maxwell, Hertz, Einstein, Planck, Hubble, Gamow, Peebles, Penzias, Wilson, and a myriad of others—and synthesized into this inescapable conclusion: *Fifteen billion years ago there was a titanic cosmic explosion that filled space with electromagnetic radiation that propagated throughout every corner of the universe. This radiation moved through the spectrum and left its weak remnants, which can be discerned by radio telescopes today. From a distant and tiny spot in that "vast, cold, universal vacuum" we are awed and humbled as we have discovered "the vanished brilliance of the origin of the worlds."*

Chapter Eleven

The Big Crunch

"And the same hour was there a great earthquake, and the tenth part of the city fell, and . . . were slain of men seven thousand . . . And the time of the dead that they should be judged . . . And when the thousand years are expired, Satan shall be loosed out of his prison, and shall go out to deceive the nations which are in the four quarters of the earth . . . Death and hell delivered up the dead which were in them; and they were judged every man according to their works."

The Book of Revelation

"We are leaving this Earth to find . . . a new dimension of truth and absolutism, far from the hypocrisies and oppression of this world."

Member, Order of the Solar Temple (1994)

When will the universe end? What will it be like? Will anything come afterward? The belief in the apocalypse, followed by survival and peace for only the virtuous, is a myth that dates as far back as the 15th century B.C., and survives fully intact today. Each year various groups predict Armageddon with the same degree of certainty that Bishop James Ussher pinpointed Creation in his writings on the Old Testament, and each year catastrophes and natural disasters are said to be precursors to the final cataclysm. "Whereas the end of the world is approaching," began royal proclamations of 10th-century England, for many believed that Christ would reign for a millennium before the day of judgment, as noted in *The Book of*

Revelation, which is an entire body of Apocalyptic prophecies. Indeed, the word "apocalypse" comes from a Greek word meaning "revelation." Today, we're reminded of the Branch Davidians in Waco, Texas, in 1993, and their predictions of chaos and destruction, and the groups that still stock their bomb shelters with provisions to survive the fire and earthquakes that will engulf the less worthy and the less resourceful.

In October 1994, a total of fifty members of the Order of the Solar Temple in Quebec and the Order of the Solar Tradition in Switzerland committed suicide. These religious sects engaged in apocalyptic prophecy that focused on an impending environmental disaster that only a select few would survive. We continue to face our destiny and deal with elements and feelings of despair that have haunted humanity in every age.

The first recorded example of apocalyptic faith is in the *Avesta*, the sacred scriptures of the Persian Empire. The *Avesta* recounts the Iranian prophet Zoroaster's vision in which he saw the Spirit of Destruction (Angra Mainyu) confront the good god (Ahura Mazda). Zoroaster lived some time between 1200 and 1500 B.C., and predicted a vast transformation and judgment of all humanity arising out of this battle in which only the just would survive.

Apocalyptic predictions are also an integral part of Christianity and Judaism. In the Gospel of Mark, Jesus overcomes the wrath of Satan to restore the communion between God and man, and in the *Book of Revelation* Jesus appears as a Zoroastrian hero who defeats Satan at the battle of Armageddon. The Old Testament reflects apocalyptic movements of early Jewish sects in the destruction of Jerusalem and the Temple by the Babylonians, and in those who sought to destroy Satan and plunge the sinners into eternal torment, while the righteous would be rewarded with eternal joy.

There are common threads throughout all apocalypse myths. We want to know about the end of the universe, and our actions are believed to bring consequences. Most myths teach that there is reward for good, punishment for bad. That there is right and wrong. Morality must have a practical impact on our human actions and interactions, or else there's less incentive to adhere to such moral principles. The myths reinforce this hope and fear—hope of eternal reward, and fear of the apocalypse, the cataclysmic end of the universe—and they attempt to add an element of meaning and urgency to life.

These ancient myths were fabricated out of fear and hope, yet they embody one element of truth and reality in our dynamic universe: *What came into existence must someday cease to exist.* People continue to fear and to be saddened by the prospect of humanity and the universe ending. Having examined the Big Bang and the formation of the universe, we'll now examine the most likely processes that could take place to cause the end of life on this Earth, the end of our solar system, and the end of the entire universe. These theories are based on observed and known scientific evidence and on the opinions of respected cosmologists and other scientists, though it's impossible to be certain which particular fate awaits us. Contrary to all apocalyptic teachings, it will happen without regard to any moral judgment and will spare no one.

Comets, Asteroids and Climate Changes Threaten Life on Earth

Routine Collisions . . . the Fate of the Dinosaurs

The viability of life on Earth, but not the Earth itself, could fall victim to a huge comet or asteroid. As mentioned in Chapter 9, there are trillions of comets trapped by the gravitational fields of the planets and the Sun traveling within our solar system. There are also thousands of asteroids in the asteroid belt between Mars and Jupiter. The great mass of Jupiter sometimes captures comets, as it did in 1994 when the comet Shoemaker-Levy 9 crashed into the planet's southern hemisphere, vaporizing clouds and creating huge plumes in its atmosphere. It was the most cataclysmic collision in recorded history. Occasionally, Jupiter also flings asteroids toward Earth or past Earth toward the Sun. It's estimated that there are about 10,000 comets and asteroids on orbits that intersect Earth's orbit and large enough (at least a half mile in diameter) to create catastrophic consequences if they were to collide with our planet. Measured by geological time, collisions are fairly routine on Earth—a large comet or asteroid strikes the Earth on the *average* of every few million years, as was the case with the comet that impacted the Earth 65 million years ago and is now widely believed to have caused the extinction of the dinosaurs.

Other possible threats to life on our fragile and tiny planet include the drastic alteration of the heat flow from the Sun caused by interference from the giant clouds of celestial gas drifting throughout the solar system, slow ecological degradation, or gradual climatic change.

The Sun's Fate Is Sealed as a Red Giant and White Dwarf

Humans Perishing in Advance

Looking at the larger picture, scientists have calculated that the entire solar system—including Earth—is certainly doomed four or five billion years down the road when our Sun nears the end of its life. The Sun is about five billion years old and can burn for another five billion years as it continues turning hydrogen into helium through fusion, the process that gives all stars heat and light. As the Sun transforms all the hydrogen in its core to helium, gravitational energy will be released, causing the Sun to expand because it will contain less mass, that is, less "self-gravity." Its surface will become somewhat cooler—to 6,700°F from its present temperature of 10,800°F—giving it a reddish hue. Ultimately, it will become a red giant, a huge fireball with two hundred times its present luminosity and about thirty times its present diameter, which will mark the beginning of the end for a star the size of our Sun. Astronomers have viewed numerous red giants in the universe, such as Aldebaran, Betelgeuse, and Arcturus. Prior to the red giant phase, the Sun's immense radiating surface will generate much greater heat even though it will contain less mass and a lower surface temperature. Human beings and all other life on Earth will have perished long before the Sun becomes a red giant, since the Earth's oceans will boil away and the atmosphere will be chemically transformed by the tremendous heat. As the fireball increases in size, it will engulf the four innermost planets (Mercury, Venus, Earth, and Mars), one by one. Then, after billions more years of erratic and violent activity, our Sun will eventually condense into a ball the size of a small planet and become an object that astronomers call a white dwarf, that is, a star of incredible density and extremely low luminosity. It will condense to such a "dwarf" because it will no longer contain enough gas pressure to support itself against its own inward gravitational pull. Ultimately, the Sun will cool, dim, and fade into the darkness of space.

Does the Universe Contain Enough Matter to Stop the Expansion and to Fall Back on Itself in a Universal Collapse?

Disintegrating Protons

Indeed, every one of the trillions of stars throughout the universe is doomed, not only because they're destined to use up their nuclear fuel, but also because they might ultimately lose the battle with *gravity*. This leads us back to the initial question: *Will the entire universe end? If so, how?* We've looked at a few scenarios regarding the fate of life on Earth, the fate of the Sun, the solar system and stars. *But what is the fate of the entire universe?* Again, the amount of matter in the universe and the resulting force of its gravity is the central factor in answering these questions. Gravity is the weakest of nature's forces when examined at the atomic level, but is the dominant force on the astronomical scale, having a range of billions of light-years and holding galaxies and clusters of galaxies together across the cosmos. If the rate of expansion of all the matter that exploded and is retreating from the Big Bang is fast enough, all such matter will be able to escape from the cumulative gravity of all the other material in the universe. That is, the expansion that began at the time of the Big Bang will continue forever. However, if the material in the universe is not expanding fast enough, it will eventually stop, and the universe will begin to contract back to a central point in a universal collapse that could be called the Big Crunch.

To envision the idea of the Big Crunch, imagine throwing a ball up in the air. Since this action does not allow the ball to escape the gravitational pull of the Earth, it returns. Rockets, on the other hand, are given sufficient escape velocity (about 25,000 miles per hour) to escape the Earth's gravity. The escape velocity that would allow an object to permanently leave the gravitational pull created by *the mass of the rest of the matter in the universe* is close to the speed of light. As mentioned in Chapter 9, it appears that the most distant galaxies in the universe are actually receding from the rest of the matter in the universe at this incredible speed—thus, fast enough to escape, never to return.

In 1992 astrophysicists at Lawrence Berkeley Laboratory at the University of California at Berkeley discovered evidence that as much as 90 percent of the matter in the universe is so-called cold dark matter, in the form of massive wisps of gas spanning two-thirds of the universe (10 billion light-years or 59 billion trillion miles across). The matter was detected by measuring its microwave radiation with instruments on NASA's Cosmic Background Explorer satellite launched in 1989, as discussed in Chapter 10. Other "missing" mass is in the form of burned-out stars. The wisps of gas and these stars are invisible to the most powerful telescopes, yet might create a strong enough gravitational field that "the ball will fall back to Earth"—that is, the universe would eventually collapse back upon itself.

Another project dedicated to finding the answer to the ultimate fate of the universe is the collection of twenty-seven radio telescopes in Socorro, New Mexico, called the Very Large Array (VLA). The VLA detects the most distant objects in the universe, thus, defining its "edge." Viewing those faint objects, we are literally looking into the past. That is, we see them as they existed billions of years ago, since it has taken their light that long to reach us. Thus, with the VLA and other instruments, we measure the volume of matter and its recession speed. Most important, we are beginning to compare the speed of recession of the distant outer objects (that is, their speed billions of years ago) with the *present* speed of recession of other objects. This information—which isn't yet fully understood—holds the key to whether the expansion is slowing down and the universe will eventually collapse (possibly leading to the next Big Bang), or whether the expansion will continue for infinity.

If the universe does collapse, clusters of galaxies will begin to mingle about a billion years before the Big Crunch. Hundreds of millions of years will elapse as the galaxies themselves merge, which will then cause immense gravitational pull on all the stars. During the final million years, all this matter will approach closely, and the temperature of the condensing universe will increase beyond that of the stars, causing them to explode. This will also cause a proliferation of cosmic bodies of extremely intense gravity, called "black holes," that will suck up all matter at an enormous rate until everything collapses into a single point of space and time with infinite temperature and density—perhaps to repeat the Big Bang all over again in a "re-creation" of the universe.

One more scenario on the fate of the universe has been proposed by a number of theoretical physicists who have picked up Einstein's work on the unified field theory. Like Einstein, they believe there exists a theory as yet undiscovered by anyone, which will explain, in one unified set or series of principles and equations, the existence and properties of all matter and energy. This is now commonly called the "grand unified theory." One aspect of most of the grand unified theories now being developed is that *all protons in the universe will fall apart in several trillion trillion trillion years.* If all protons are destined to disintegrate on their own over the next several trillion trillion trillion years, as the theory predicts, every atom will also fall apart. Physicists from the University of California at Irvine, the University of Michigan, Brookhaven National Laboratory in Upton, New York, and other locations around the world are attempting to prove or disprove that theory in the hope of better understanding subatomic particles and tying them to all physical phenomena in the ultimate theory in physics.

As we approach the year 2000, the Millennium Watch Institute, which tracks apocalyptic prophecies, reported that about one thousand organizations and individuals in the United States alone publish books and articles about global transformations that will lead to the apocalypse. Some warn of floods and earthquakes, while others predict war or famine. Though it lacks the long and rich history and tradition of these myths, the truth about our fate and the fate of the entire universe lies in asteroids, comets, the Sun's fusion process, the cumulative gravity of the universe, and the theory of proton disintegration.

Looking back on the discoveries of Newton, Rutherford, Bohr, Einstein, Hubble, and the other scientists discussed in these pages, we see that each successive step in acquiring scientific knowledge has given us a new perspective not only on the physical universe, but also on the human race. As we've moved from the pre-Copernican center of a finite and changeless universe of 5,000 stars to the post-Hubble fringes of a galaxy of 300 billion stars amid billions of other galaxies, we've kept testing the limits of our knowledge. Relative to the universe as we now know it, our entire world is smaller than a grain of sand among the other grains on all the world's beaches.

The universe began as an empty space filled with elementary particles, and we're not sure how it will end. Between this beginning and end, humanity has arisen, with its diverse and fascinating history and culture. As we continue to apply our reasoning ability to the understanding of the physical world, and until life is actually found elsewhere in the universe, we humans appear all the more incredible. We are more aware of the context in which we exist. We inhabit a brief interval of time with barren, lifeless planets as our companions, all a result of the Big Bang. We exist in a remote corner of this galaxy in which the next closest star is *23 trillion miles away*. Though people have found comfort throughout history in the rich tradition of myth, there is greater comfort in what we now know, for it is rich in truth.

Now that we've explored the major discoveries in physics and astronomy, we will examine directly and in a similar depth this *matter that lives*—the unique and magnificent configurations and beings that we call life. We will explore the great mysteries of life and the intricate ties between the physical universe and the living universe that arose on Earth. We will see the relationship between the living material and the chemical elements of the periodic table. In Parts Five, Six, and Seven, we will place life forms in their proper context amid the atoms of which they are constructed and the physical forces and environment that have molded and determined life's creation and destiny. We will carefully identify the conditions, steps, and developments that have occurred over billions of years as the elements present in Earth's primordial soup first coalesced into life, diverged a millionfold, eventually climbed onto dry land, and, in the case of one such form of life, evolved to the point where it was able to squint and look back deeply into the fog of time and discern its own history.

Part Five

Evolution and the Principle of Natural Selection

In Focus

Of the seven greatest scientific discoveries, three stand out as completely changing our fundamental concepts about life and the universe: (1) gravity and the physics of the universe; (2) the Big Bang; and (3) Darwin's theory of evolution, which we will explore in the following pages. Evolution is far from a theory—it is a principle beyond question and beyond debate. Nevertheless, with their destructive and irrational agenda, 20th-century Creationists, straight out of the Dark Ages, seek to place emotion before reason and make evolution a debatable issue. The truth is, all forms of life share a common ancestor, and through a process that naturally "selects" the characteristics important for survival, each species, including humans, gradually adapts to its changing environment.

In the 140 years since Darwin wrote *The Origin of Species,* we've made astounding advances in understanding how evolution works, including 20th-century fossil discoveries of *Homo sapiens'* predecessors and the precise mechanisms of evolution revealed in the cell and DNA, which we will discuss in Parts Six and Seven. Also, a number of unexpected discoveries in other fields have filled in the picture. For example, in Chapter 15 we'll look at the role of plate tectonics in changing environmental conditions, which in turn causes natural selection to occur.

Charles Darwin's discovery approaches Newton's in significance and is similar in the sense that both men developed their propositions with minimal reliance on their predecessors. Their work was largely individual, incredibly original, and comprehensive. Darwin has been described as "the Newton of biology," for Newton opened the door to our understanding of the universe, and Darwin opened the door to our understanding of life.

Chapter Twelve

The Rocks of Genesis

"In the beginning God created the heaven and the earth . . . And God created . . . every living creature that moveth . . . And the Lord God formed man of the dust of the ground, and breathed into his nostrils the breath of life; and man became a living soul."

The Book of Genesis

"The belief that species were immutable productions was almost unavoidable as long as the history of the world was thought to be of short duration . . . The chief cause of our . . . unwillingness to admit that one species has given birth to clear and distinct species is that we are always slow in admitting great changes of which we do not see steps."

CHARLES DARWIN
The Origin of Species (1859)

We now turn from the telescope and distant heavens and begin to look at the origin and nature of life on this Earth. As dangerous and controversial as Galileo's and Newton's discoveries were thought to be, those who would dare challenge fundamental beliefs about life itself, particularly about mankind, would find themselves in the midst of even greater danger and controversy. That should have been expected—the origin of life is a much more personal and more emotional subject than the Earth's location or the nature of gravity. But, unfazed by their many detractors, a new group of scientists emerged in the 19th century and began a revolution in geology and biology that resulted in an understanding of the age of the Earth and the nature and origin of living things. This new understanding was so profound it removed mankind's childlike innocence forever.

Spontaneous Generation of Life Is Disproven But There Is No Reason to Question Creation

In God's Image

Until the Renaissance, reproduction was believed to be a result of a supernatural event, permanently beyond the descriptive powers of science, the same as the creation of life in Genesis. Reproduction of the simpler forms of life was thought to occur spontaneously out of nonliving matter, such as the sudden appearance of maggots from meat, and the generation of worms from trees, beetles from dung, and mice from garbage. This set of beliefs was called "spontaneous generation." Another notion common in academic circles of the 17th century was the preformation theory, which held that the sperm or the egg (depending on which version one supported) contained microscopic preformed beings created for all future generations at the time of Adam and Eve, awaiting fertilization to trigger a mysterious growth mechanism. The Italian physician and biologist Marcello Malpighi (1628–1694) popularized this theory, though his decades of work as the founder of microscopic anatomy and other real scientific advances in embryology, anatomy, and pathology far outweigh any negative impact on advances in biology resulting from his belief in the preformation theory.

The British physiologist William Harvey (1578–1657) was the first to demonstrate that spontaneous generation was impossible because every animal comes from an egg, and the Italian biologist Francesco Redi (1626–1697) conducted experiments showing that the presence of maggots in putrefying meat is not caused spontaneously, but from microscopic eggs laid on the meat by flies. Through his decades of research on bacteria, protozoa, and other simple forms of microscopic animal life, the famous Dutch lens-maker, Antoni van Leeuwenhoek (1632–1723), who made great improvements to the microscope, contributed much to dispelling the idea of spontaneous generation. However, he also promoted the preformation theory. In the late 1700's, the Italian priest and physiologist Lazzaro Spallanzani (1729–1799) conducted a series of experiments that proved that actual contact between egg and sperm is essential for the development of new animals.

Biological explanations for reproduction were gradually recognized and accepted during the 1700's. Yet the entire field of biology—the understanding of all living things—remained in its infant stages. Until about 1830, the common and accepted view of the world was that we lived on a stable and unchanging planet populated by immutable, unchanging species, just as God created them, and that He created Earth itself to be a home for man. All plants and animals were believed to be eternal in their present form—an understandable view, for the theory of a changing universe with a beginning and end was not developed until the 20th century, with the discovery of the Big Bang and the recession of the galaxies, as we saw in Part Four. Aristotle's crystal spheres had been shattered well before 1830, and as a result of centuries of exploration, people had accepted the idea of "Antipodes" (a Southern Hemisphere with places geographically diametrically opposed to the centers of Western civilization). However, merely because the Earth rotated and revolved, and simply because Newtonian physics applied equally to New Zealand and Great Britain, did not mean Earth itself or the life on it had physically changed over time.

Except for the Earth's location in the solar system, the Bible was still taken literally. The universe was created in six days. As mentioned in Chapter 9, Archbishop James Ussher of Ireland calculated the year of creation to be 4004 B.C. and the *moment* of creation to be 2:30 p.m. on Sunday, October 23. This belief made human existence clear and comprehensible. People drew comfort from their feeling that God had not cruelly withheld our own past from us—that we hadn't missed anything since the beginning of the universe because there was no universe, no history, before the Bible. The word "prehistoric" did not enter European vocabularies until the mid-19th century. Using 50 years as an average lifetime,

5,800 years is only 116 consecutive lifetimes. Thus, mankind had emerged from an unknown darkness 116 lifetimes ago, much like emerging from the womb. We believed ourselves to be children of God, innocent and protected, and we had no basis to doubt or question that belief. The familiar idea of birth was adapted to the concept of Creation. God created Adam and gave birth to mankind, as depicted in Michelangelo's painting on the ceiling of the Sistine Chapel in Rome. The art and literature of Western civilization embedded this image deep within our culture. It would take another revolution to wrench that image from our minds.

Categorizing and Comparing Living Things Leads to the Question of Whether Life Forms Are Permanently Fixed

The Sixth Day of Creation

Aristotle had taught that everything in nature seeks perfection, reaches for perfection, with some achieving (like humans), and others failing (like all lesser animals), but that living things do not evolve in the biological sense. His elaborate "Scale of Being" philosophy was a living ladder that placed minerals at the bottom and man at the top, with nothing changing. Species were eternally fixed signposts identifying the levels of perfection or imperfection, having nothing to do with their anatomy or actual evolutionary relationship. It was this concept, coupled with Genesis, Noah's ark, biblically described humans, and the power of religion, that directed rational thought away from any idea of evolution and allowed Aristotle's fixed-species concept to outlive his fixed-Earth concept by two centuries.

In 1691 the British naturalist and botanist John Ray (1627–1705) wrote *The Wisdom of God Manifested in the Works of Creation*, which described a pattern of living things based on their structural characteristics and internal anatomy. This work and others perpetuated and refined Aristotle's concept by logically organizing, identifying, and categorizing the living things known at that time and giving them a nomenclature, which rigidly fixed the immutability concept. At the same time, however, it had the paradoxical effect of laying the groundwork for comparative anatomy, which later became an important part of evolutionary theory. That is, evolutionary theory first needed this clear picture of anatomical similarities and differences of species on which to build.

The most significant figure in the trend to classify, describe, and compare was Carl Linnaeus (1707–1778) of Sweden. The great modern anthropologist and philosopher Loren Eiseley described Linnaeus's love of the incredible variety of life as a "poetic hunger of the mind to experience every leaf, flower, and bird . . . He was the naming genius par excellence . . ." In *Systema Naturae*, first published in 1735, Linnaeus created a clear and efficient method of naming all animals and plants. Each was identified by two names. The first was generic to identify a group (or genus) of visibly related creatures, such as all doglike forms. The second referred to a more specific group, a species, such as wolves. Thus, the wolf is *Canis lupus*. In addition to the genus and species reflected in the name, the Linnaean system (as modified and refined by modern biology) has five increasingly more specific levels, each with additional categories. Thus, the seven levels of all living things, beginning with the largest and most general, consist of:

Kingdom: Animals, plants, fungi, protozoans, bacteria.

Phylum: For example, there are twenty phyla in the animal kingdom. One is Chordata—all animals with head-to-tail nerve cords (fish, snakes, bats, humans, etc.).

Class: For example, one class of chordates is Mammalia, which are animals that have mammary glands to nurse their young, as well as other specific characteristics (bats, whales, humans, etc.).

Order: For example, one order of mammals is Primates, which possess all of the characteristics described above for mammals and chordates, and also have five fingers and toes, nails instead of claws, hands that can grasp, thumbs, and a relatively large brain (chimpanzees, apes, orangutans, humans, etc.) Humans share the same kingdom, phylum, class, and order with apes.

Family: For example, one family of primates is Hominidae (which are also mammals and chordates), characterized by their ability to walk on two legs. There are at least three extinct species of hominids—*Homo erectus, Homo habilis,* and *Australopithecus* (which will be discussed in Chapter 14)—and one remaining: *Homo sapiens.*

Genus: For example, one genus of hominid is *Homo* (from the Latin, meaning "human," not the Greek, meaning "same")—characterized by larger brain size and a head shaped like a modern human's. Again, *Homo sapiens* is the only living species filling the genus *Homo.*

Species: Of the genus *Homo* and family Hominidae, *Homo sapiens* (sapien meaning "wise") is the only species. Taking another example, the Blackburnian warbler (*Dendroica fusca*) is one species that shares the genus *Dendroica* with 27 other species, and the family Parulidae with 125 other species. In modern terms, *a species means all the members of a group that are genetically capable of interbreeding and producing fertile offspring.*

Linnaeus's system was extremely useful and timely during the 18th and 19th century, with the increasing number of sea voyages that resulted in the rapid discovery of thousands of new species of plants and animals in the New World, Australia, Africa, and the Far East. Linnaeus rose to fame as a result of his mastery of the living world and the order he imposed on it. Largely because of his development of this system, naturalists were given official posts on board overseas voyages, such as the one which was to set Charles Darwin on the course of evolutionary theory.

In the early editions of *Systema,* Linnaeus maintained the idea of the fixity of species—that modern species were as fixed as they had been on the sixth day of Creation. His system of class and gradation made it more clear than ever that each living thing possessed a specific form. However, as Linnaeus continued his lifelong work, the paradox began to surface as he found it increasingly difficult to cling to his original thesis of the immutability of species. Recognizing the results of crossbreeding farm animals and the existence of small inherited changes called mutations, he became confused as to "whether all these species are the children of time, or whether the Creator from the very beginning of the world had restricted this course of development to a definite number of species." With this unanswered question in mind, he removed from later editions of *Systema* his assertion that all species are fixed and no new ones can arise. So Linnaeus's system was not only a basis for comparative anatomy of living things, but one that could reflect change in those living things.

Modern Geology Is Stillborn as Creationism Persists

Time Is of the Essence . . . Mammoth Elephants and Saber-Toothed Tigers

As early as 560 B.C., the Greek philosopher Xenophanes of Colophon (570–475 B.C.) found seashells embedded in the rock strata high in the mountains of Greece. He believed that the seashells became embedded there during a catastrophic flood. In the mid-1600's, the French naturalist Isaac de la Peyrere (1594–1676) published a book about some strangely shaped stones that had been collected in the French countryside, and suggested that the stones had been shaped by primitive men who lived before the time of Adam. De la Peyrere's life was threatened and his book was burned publicly in 1655.

Then in 1749, the French naturalist Georges-Louis Leclerc de Buffon (1707–1788) published the first book of his 44-volume work on natural history. This monumental study reflected Buffon's revolutionary ideas in geology and biology, and was the first work to treat those subjects scientifically. He concluded that some animal life had become extinct and discreetly suggested that animals might have undergone some type of evolutionary change. He also stated that some mammals might have *common* ancestors that were physically different from those mammals of his day. Scientists and theologians were critical of Buffon's ideas of evolution, his concept of geological history in stages, and his suggestion of a time scale of up to 35,000 years to account for such stratification and the history of life on Earth. Just as the Catholic Church had forced Galileo to recant in 1633, the theology committee at the University of Paris forced Buffon to retract in writing certain passages to avoid censure: "I abandon whatever in my book concerns the formation of the Earth," said Buffon, "and all that might be contrary to the narration of Moses."

In 1771, Johann Friedrich Esper (1732–1781) found human bones in a cave in Germany along with the skeleton of an extinct bear. Near Hoxne, England, in 1790, the British archaeologist John Frere (1740–1807) found some Stone Age flint implements among the bones of other animals that no longer existed on Earth. In other locations in Europe, human skulls were found with remains like these and with numerous other types of fossil discoveries—all forming an image not found in the Bible, an image of an Earth that was once inhabited by mammoth elephants, woolly rhinos, saber-toothed tigers, and many other species, all now extinct.

However, apart from the need for fossil evidence on the *biological* aspects of such a revolutionary theory as evolution, there first needed to be evidence of sufficient *time* for it to occur. Buffon had speculated on this point, but the crucial evidence was locked in the Earth itself, waiting to be liberated through careful observation and analysis by those who were to develop a completely new field of science. After years of such observation and research, James Hutton (1726–1797) of Scotland presented the first well-documented analysis on the age of the Earth to the Royal Society of Edinburgh in 1785, in a paper entitled "Theory of the Earth." Hutton described a compelling picture: *Soils form by the weathering of rocks; tides and the pounding of waves erode the coast; layers of sediment accumulate; and the general cycles of sedimentation, uplift of hills and mountains, and erosion can be seen everywhere.* With this paper, Hutton created a new science, geology, out of nothing, and became the father of modern geology. He originated the fundamental principle on which geology is now based, called uniformitarianism or uniformity, which holds that the rocks and other inorganic materials of the Earth are formed and modified by a continuous and generally uniform series of natural phenomena, such as rain, wind, tides, and gradual shifts in the Earth's crust. The impact of catastrophes such as floods and earthquakes was recognized by Hutton, but as exceptions to the basic geologic processes of uniformitarianism. "The present is the key to the past," said Hutton, meaning that all the conditions responsible for observed (past) geological phenomena are the same conditions that exist today. That is, past phenomena can be fully explained by current processes *plus time*.

Ten years after his original presentation to the Royal Society, Hutton compiled a more comprehensive analysis in a two-volume work, *Theory of the Earth*. He squarely proposed that the Earth was at least hundreds of thousands of years old. "James Hutton brooding over a little Scottish brook that carried sediment down to the sea," wrote Loren Eiseley, "felt the weight of the solid continent slide uneasily beneath his feet and cities and empires flow away as insubstantially as a summer cloud. . . . He discovered an intangible thing against which the human mind had long armored itself . . . time." Hutton opened this expanded window of time, allowing others the opportunity to place the growing number of fossil discoveries in their proper context.

Despite the force and persuasiveness of Hutton's lectures and writings, many scientists and others challenged uniformitarianism. They would not accept any concept that conflicted with the popularly accepted view called catastrophism, which held that geologic formations were caused by sudden catastrophic upheavals and events, such as Noah's flood, and that the Earth was created by God about 5,800 years earlier to be a home for man. Hutton was branded a heretic by many, but to a growing number of scientists it became apparent that catastrophism would have to be abandoned as a viable principle or modified to incorporate Hutton's theories, while somehow preserving the tenets of Christian theology.

In 1796 the French zoologist and statesman Georges Cuvier (1769–1832) began to accumulate his own evidence of the Earth's history. While a professor of natural history at the Collège de France, Cuvier discovered woolly mammoth elephant bones near Paris, giant salamanders, flying reptiles, and other extinct species. His fossils indicated a correlation between the rock strata and the development or changes in extinct species—that is, the deeper the strata, the less similar to animals now living, as compared to the fossils in the more recent strata. With the help of mineralogists, Cuvier tried to figure out how the bones of these fascinating extinct creatures had become entombed in the ground. He more fully developed the new scientific discipline of comparative anatomy (building on Linnaeus's foundation) through his brilliant writings, such as *Lessons in Comparative Anatomy* (1800), in which he stated his principle of the "correlation of parts," and *Research on the Bones of Fossil Vertebrates* (1812).

It would seem that Cuvier had no choice but to confirm Hutton's theories, as the Earth revealed its ancient story told by layer upon layer of seashells, bones, different types of rock, the evidence of lakes and seas, and the recession of water followed by a new marine era. But as a devoutly religious man, Cuvier saw his task to be reconciling the evidence of these phases of geological events with an Earth less than 6,000 years old, placing him in the difficult role of becoming the leading spokesman for catastrophism. He concluded that the popular concept of catastrophism was right and that James Hutton was wrong. A series of vast floods, said Cuvier, had occurred that accounted for the sedimentation and the extinction of species. He expounded on his version of catastrophism beginning in 1812, and wrote supporting treatises in 1815, with *Theory of the Earth*, and in 1825, *Discourse on the Revolutions of the Globe*, despite the increasing support for Hutton in a scattering of new writings, such as *The Stratigraphical System of Organized Fossils*, published in 1817 by William Smith (1769–1839). Cuvier maintained that the species spared from the sudden land upheavals and floods repopulated the Earth. He believed new species weren't actually new, but were from unexplored parts of the world. Cuvier's unfortunate legacy was to create a model for 20th-century Creationists.

Without a realistic concept of geologic time, catastrophism remained the popular view, preserving what were to be Aristotle's last gasps in biology, and perpetuating the divine origin and immutability of species.

Geologic Time Becomes a Reality But Evolutionary Theory Remains Weak and Unsupported

In the Image of an Ape

The strongest support for a theory of evolution prior to the 19th century was developed through the research and writings of Erasmus Darwin (1731–1802) (Charles Darwin's grandfather) and Jean-Baptiste Lamarck (1744–1829). Erasmus Darwin was a prominent doctor in England and was also known as an innovative and radical thinker. His version of evolutionary theory held that species modify themselves by adapting to their environment through a conscious effort of some kind, which came to be known as the doctrine of acquired characteristics.

Jean-Baptiste Lamarck, a contemporary of Erasmus Darwin, was the better known of the two, and maintained a similar evolutionary philosophy. Lamarck was a French botanist and biologist, and made great contributions to the proper classifications of plant and animal life (class, order, genus, etc). In his 1809 publication, *Zoological Philosophy*, he set forth two principles that he believed explained the vast array and development of life forms: Organs improve with use and are weakened by disuse; and such changes are preserved in animals and passed on to their offspring. So, for example, he believed that giraffes have long necks because they stretch to reach leaves in trees. This was sort of a "reverse Darwinism" in that the creature controls its own destiny, rather than the environment creating the creature. It was conjecture, not based on any scientific approach or proof.

Lamarck's version of the doctrine of acquired characteristics was a combination of his own view of God's plan and Aristotle's Scale of Being. Lamarck wrote of life ascending toward higher levels to perfect the Creation. Although he recognized that changes in the physical environment cause changes in the needs of living things, he blamed these environmental factors for *interfering* with all animals' abilities to achieve perfection in accordance with God's ultimate purpose. Like other 18th-century evolutionists, he viewed variations as being restricted within limits. For example, a gray pigeon and a white pigeon might share a common ancestor from many generations before, but the dog, wolf, and bear surely could not. Lamarck's general views were interesting and controversial and won support from some biologists in the early part of the 19th century. Yet his inability and refusal to support his theories with specific data led to his eventual ostracism. He died blind and impoverished.

So, by the 1820's no one had yet seen the full view through the window opened by Hutton. Geology and biology were not yet linked in any cohesive manner: Linnaeus had created the system of categorizing living things into seven levels; we had begun to view cells with the microscope and ask about the developmental stages of living things; Hutton had made a strong case for uniformitarianism, but Cuvier's catastrophism blinded people from a view that allowed for the necessary geologic time; Buffon, Erasmus Darwin, and Lamarck were known in intellectual circles as suggesting that animals might not be as immutable as once thought, but the process and driving force behind evolution remained vague and the subject of conjecture and speculation. The suggestions of evolution and an ancient Earth were still faint voices drowned out by the popular views of fixity of species and catastrophism.

In the wake of Cuvier's popularity in Europe in the early 1800's, an English geologist named Charles Lyell (1797–1875) examined rock formations and fossils in greater detail than had ever been done before. He published *Principles of Geology* in 1830, in which he adopted Hutton's views and more convincingly demonstrated that despite Cuvier's pronouncements, catastrophism simply didn't make sense. Even though catastrophes occur from time to time, the natural forces of wind, water, and shifting earth have generally operated slowly and uniformly.

Lyell supported his principles with an unprecedented number of examples that he obtained through his extensive travels around the world. Faced with this overwhelming evidence, people finally began to accept the concept of a slow, uniform geological process. For the first time in history, it appeared the Earth might even be *millions* of years old. Indeed, perhaps the Earth had changed. Perhaps inorganic formations of rocks, beaches, and mountains had eroded and developed entirely new configurations. In geology, the tide began to turn in favor of Hutton, Lyell, and the young naturalists who were not mired in the past.

Nevertheless, even if the Earth was older than originally thought and had undergone change as described by Lyell, it remained unacceptable—scientifically, philosophically, and religiously—to suggest that *life itself* had somehow evolved. Therefore, in 1830, when the Belgian paleontologist Philippe-Charles Schmerling (1791–1836) found stone artifacts and two human skulls together with extinct rhinoceroses and mammoths in Belgium, the proximity was explained away as coincidence or fraud. In 1838 the French archaeologist and writer Jacques Boucher de Perthes (1788–1868) found a large number of flint hand axes and other stone objects near Abbeville in northern France that appeared to be ancient and handcrafted. Boucher de Perthes presented his finds to the scientific community with his theory that the hand axes were made by primitive men who existed prior to the deluge described in Genesis. Despite this and other mounting evidence, Boucher de Perthes's theories and writings were ignored, much as Isaac de la Peyrere's similar discoveries had been rejected in the 1600's, because without radioactive dating techniques or the ability to chemically analyze these materials, there was no scientific method of proving their age.

Thus, although Lyell's geologic principles continued to gain acceptance and the door slowly opened to question the *age* of Earth and its living things, this didn't connote evolution. It meant only that the immutable life forms created by God might have been in existence for longer than 5,800 years. It was one thing to accept a changing Earth and an expanded view of biological time, but the idea of changing life forms remained outrageous and the province of a few fanciful philosophers. The view stated by John Ray endured: "Whatever may be said for ye Antiquity of the Earth itself and bodies lodged in it, ye race of mankind is new."

Amid growing evidence of primitive humans, extinct species with human artifacts, the complete lack of fossils of existing species, and the work of Buffon, Lamarck, and Erasmus Darwin, the question persisted: *Do living things evolve?* If the inorganic world changed over time, as Hutton and Lyell had shown, is it possible the organic world also evolved over time? Why the extinct animals? The human bones in caves? Was man once as primitive as the tools and bones suggest? *Was man ever less than man?* Questions of explosive proportions.

As long as species were fixed in their present form and the Earth's changes could be attributed to sudden and cataclysmic causes, such as Noah's flood, there was peace between the theologians and scientists. That had been the state of affairs during the first few decades of the 19th century—an unspoken truce, for the geologists and biologists had gone as far as possible to reconcile their observations with the Scriptures. But Hutton's and Lyell's changing Earth view, coupled with attempts at evolutionary theory, marked the beginning of the end of this uneasy compromise.

Against God's word, Copernicus, Kepler, Galileo, and Newton had taken mankind out of the center of the universe. Now another heretic was about to peer through the window opened by Hutton and Lyell, and describe an unimaginable vista. In describing his view, he would offend humanity further by showing that Michelangelo's painting on the ceiling of the Sistine Chapel in Rome must be totally redesigned, for humans were *not* created in the image and likeness of God, and the fingertip endowing Adam with existence was not God's at all. *It was the fingertip of an ape.*

Chapter Thirteen

The Indelible Stamp of Our Origin

"The great synthesizer who alters the outlook of a generation, who suddenly produces a kaleidoscopic change in our vision of the world, is apt to be the most envied, feared, and hated man among his contemporaries."

LOREN EISELEY
Darwin's Century (1958)

Indeed, the boldest and most controversial revelation of the 19th century and perhaps all of history was the theory of evolution by natural selection developed by Charles Robert Darwin (1809–1882). It fired the emotions of Darwin's generation and each one that followed—including those who hailed his genius, and those who vowed to burn him at the stake.

Darwin Sets Sail on the *Beagle*

Finches and Ostriches Offer a Clue

Darwin's mother died when he was eight, and his eldest sister, Caroline, raised him. He attended exclusive private schools as a boy, but was not a good student, caring "for nothing but shooting, dogs and rat-catching," according to his father, a wealthy doctor. Reminiscent of Einstein's father's comment, the elder Darwin once told his son, "You will be a disgrace to yourself and your family." After dropping out of medical school at Edinburgh University, Darwin graduated as a theology student from Christ's College at the University of Cambridge, England, in the class of 1831, and intended to have a life of service in the Church of England. But there was no hurry about the ordination, since there was usually a wait of a year or two before a deaconship or other position was available. Darwin's father agreed to support him in the interim.

He'd done well in his naturalist classes in college, and considered taking a "geologizing" trip or a voyage on a merchant vessel to another part of the world to expand his horizons before making the lifelong commitment to the clergy. He admired his botany professor, John Stevens Henslow (1796–1861), who'd made botany and biology popular subjects at the university and with whom Darwin had become friends, but he had never considered basing his life or profession on those subjects. However, in late August 1831, Darwin received a letter from a government official named George Peacock that led to a change in Darwin's views on the importance of biology, and would ultimately change the entire course of history:

My Dear Sir,

Captain FitzRoy . . . sails at the end of September in a ship to survey . . . the S. Coast of Terra del Fuego, afterwards to visit the South Sea Islands & to return by the Indian Archipelago to England. The expedition is entirely for scientific purposes & the ship will generally wait your leisure for researches in natural history . . . The Admiralty are not disposed to give a salary, though they will furnish you with an official appointment . . . If a salary should be required however I am inclined to think it would be granted . . .

The recommendation for Darwin's appointment as official naturalist on the British Navy surveying ship *H.M.S. Beagle* came from Professor Henslow. It was a unique opportunity—a voyage around the world collecting and identifying every example of living plant and animal encountered. That was the extent of botany and biology as it existed at that time—categorizing living things and fitting them into the system established by Carl Linnaeus. The trip was to last no longer than two years, and then Darwin planned to return to the Church.

He was selected despite being the *third* choice as naturalist on the *Beagle*. Leonard Jenyns (Henslow's brother-in-law) had been the first to be offered the position. Professor Henslow himself was the second. Jenyns declined because of his commitment to his church congregations, and Henslow declined because of his wife's objections. Darwin came close to being left on shore because of a personality conflict and differing political philosophy with the ship's temperamental Captain Robert FitzRoy (1805–1865). Also, Darwin's father initially objected to the undertaking, but eventually relented after lobbying by Darwin's uncle, Josiah Wedgwood, the famous English pottery designer and manufacturer.

After all this and a series of delays because of bad weather, the *Beagle* weighed anchor at 11 a.m., December 27, 1831, manned by a small crew. Charles Darwin was crammed into his living quarters in the poop cabin with barely enough headroom to straighten his six-foot frame, four inches on either side of him when upright, five feet of hammock, and room for few possessions. The ship took Darwin and the crew from Plymouth, England, to the Canary Islands, down the east coast of South America, up the west coast to the Galapagos Islands, then to New Zealand, Australia, Tasmania, Madagascar, around the Cape of Good Hope, to the Ascension Islands, back to Brazil, and finally home.

His singular purpose during the voyage was to collect samples of rocks, fossils, sea animals, plant life, and animals on shore near the harbor towns where they stopped during their travels. His responsibility as naturalist included proper preparation and preservation of the specimens, and packaging and shipping them back to Professor Henslow for observation, categorization, and analysis. As Darwin carefully observed and examined all these life forms, extinct and living, and the surrounding geologic formations, he began to develop an insight into the relationship between that changing geology and the life it sustains. He was particularly influenced by Lyell's *Principles of Geology*, a going-away gift from Henslow, and saw firsthand the cycles Lyell wrote about: the wear and tear of waves, the fall of forest and rock, gradual rising of land, and the erosion caused by rivers.

Contrary to common belief, Darwin did not develop his theory of evolution during the thirty-six days he spent on the Galapagos

Figure 13-1

The Voyage of the *Beagle*, 1831–1836

Plymouth
sailed Dec.27, 1831

Falmouth
landed Oct. 2, 1836

Off Tenerife
Jan 6, 1832
unable to land because
of quarantine

Azores
landed Sept. 20, 1836

Cape Verde Is.
landed Jan.16, 1832
landed Aug.31, 1836

Ascension Is.
landed July 20, 1836

Bahia landed
Aug 1, 1836

St. Helena
landed July 8, 1836

Cape Town
landed June 1, 1836

Cocos Keeling Is.
landed April 2, 1836

Mauritius
landed Apr. 30, 1836

King George's Sound
landed Mar. 7, 1836

Sydney
landed Jan. 12, 1836

New
Zealand
landed
Dec. 21, 1835

Hobart
landed Feb. 5, 1836

Fernando de Noronha
landed Feb 20, 1832

landed
Feb. 28,1832

Rio de Janeiro
landed
Apr.5, 1832

Montevideo
landed
July 28, 1832

Falkland Is.
landed Mar. 1, 1833
landed Mar. 10, 1834

Passed out of
Strait of Magellan
June 10, 1834

Valparaiso
landed July 23, 1834

Callao
landed July 20, 1835

Galapagos Is.
landed Sept. 16, 1835
sailed Oct. 20,1835

Tahiti
landed Nov. 15, 1835

Islands, while the *Beagle* charted its shore during the fall of 1835. But Galapagos did provide an essential piece of the puzzle by raising questions about the great variations he saw in the indigenous animal life there. Darwin studied the finches he caught on two different islands and found they had differently shaped beaks. Similarly, he observed that shells of the tortoises of the various islands differed in shape, color, and thickness. He also noted differences in the tortoises' size, and the length of their necks and legs. Distinct physical characteristics of these and several other animal species on the sixteen islands prompted Darwin to speculate about what causes the differences. Had each of the islands developed its own species? His experience in the Galapagos also led him to broader questions: Why are animals similar in places that are so very different? Why is the South African ostrich so much like the South American rhea? But then, why aren't the *fossil* animals found in the Galapagos identical to the *living* forms there? The voyage gave rise to these and other questions and to doubts about the literal truth of Genesis. By the end of this five-year awakening, Darwin began to scrutinize every specimen he encountered and to establish a revolutionary new paradigm for living things.

The voyage that was supposed to take two years ended up taking five because of bad weather, mechanical problems, the need to resurvey where errors had been made, and all the unexpected events of a trip on the high seas in a small vessel in the early 1800's. By the time the *Beagle* docked at Falmouth, England, on October 2, 1836, Darwin had already gained some notoriety in England for the vast collection of specimens he'd gathered and sent to Henslow during the voyage. Soon after his return, he began presenting a series of papers before various prestigious scientific groups such as the London Geological Society on subjects ranging from fossil bones he'd dug out of the cliffs in South America to his theory of how corals are formed. Fascinated by the immense variety of species, their interrelationship, and the questions about their history, Darwin abandoned his earlier course as a man of the cloth and set a new one in nature.

The Theory of Evolution Is Born

One Common Ancestor

In mid-July 1837, Darwin began writing in his journal on the subject that had begun to take form on the Galapagos Islands: *the transmutation of species.* He usually avoided using the word *evolution* and preferred *transmutation,* since he wanted to emphasize that species of plants and animals change in response to their environment but might remain that species. He later described how he came to address the issue:

> When on board *H.M.S. Beagle,* as naturalist, I was much struck with certain facts in the distribution of the organic beings inhabiting South America, and in the geological relations of the present to the past inhabitants of that continent. These facts ... seemed to throw some light on the origin of species ... On my return home, it occurred to me, in 1837, that something might perhaps be made out on this question by patiently accumulating and reflecting on all sorts of facts which could possibly have any bearing on it.

At about the same time, Darwin happened to be reading *An Essay on the Principle of Population,* by the famous English economist and demographer Thomas Malthus (1766–1834), in which he maintained that human population and demand will always exceed the production of food and other needs. Darwin described Malthus's views in this way:

> There is a constant tendency in all animated life to increase beyond the nourishment prepared for it ... Nature has scattered the seeds of life abroad with the most profuse and liberal hand; but has been comparatively sparing in the room and the nourishment necessary to rear them ... Population has this constant tendency to increase beyond the means of subsistence ...

In the course of reading Malthus in October 1838, he experienced a great revelation:

> Being well prepared to appreciate the struggle for existence which everywhere goes on ... it at once struck me that under these cir-

cumstances favorable variations would tend to be preserved, and un-favorable ones to be destroyed. The result of this would be the formation of new species. Here then I had at last got a theory by which to work.

Darwin took Malthus's observation and applied it to his own observations of plant and animal species. He recognized that the reproductive potential of plants and animals vastly exceeds the rate necessary to maintain the population of these plants and animals at a constant level, yet such population sizes remain fairly constant. Based on this, Darwin concluded that the plants and animals that survive the intense competition among all life must be better equipped to live in their particular environment than the ones that do not survive. Predators battle for territory and prey, grazing animals seek the prime lands, plant life competes for space in which to expand and proliferate, males of many species fight for the right to the females. The swiftest, strongest, and smartest prevail, and those vital characteristics are inherited by their descendants. Natural selection allows the changes that enhance survival to be passed on, and eliminates changes that do not. Though the science of genetics did not yet exist, Darwin surmised that essential qualities of the surviving predator, grazer, plant, and victor for the female were *somehow* passed on in their seed: "All these results . . . follow from the struggle for life. Owing to this struggle, variations . . . profitable to the individuals of a species . . . will tend to the preservation of such individuals and will generally be inherited by the offspring . . ."

That was his proposition, his theory of evolution: *The characteristics that contribute to each animal's superiority and success are preserved and passed on to future generations, each species gradually adapting and continuing to do so as the environment changes, in terms of climate, geography, available food, predators' adaptations, and all other aspects of that living thing's surroundings.* His answer to the question he initially posed on Galapagos was brilliant in its simplicity: Plant and animal species undergo change because the most fit members of a species survive the gradually changing environmental elements. They adapt. They evolve along with their surroundings. If they fail to do so, they die, and if enough die, they eventually become extinct. Therefore, most extinct species are simply the earlier form of their progeny that flourish today. Though there are species that died off in massive numbers, like the dinosaurs, most fossils represent extinct species that did not actually abruptly disappear or die off without successors or offspring—they simply evolved over millennia into later and "better" models.

Yet, there are presently about one million species of animals (including 751,000 species of insects)—a million later and better models, not just one. So, if individual members with favorable naturally selected traits are free to breed with all members of the species, how did natural selection originally result in this great proliferation of species? The answer lies in "speciation"—the ability of a species to separate and subdivide into other species—and particularly one aspect of speciation called "reproductive isolation." Reproductive isolation can occur in a number of ways, usually because of geographical or physical developments that result in barriers or distance between members of the same species: rivers change course, mountains form, and continents drift apart. Also, groups of individuals in a species often strike out into new areas on their own in search of a better food supply or reproductive environment. Any of these developments can result in a degree of isolation that will take the organism and group down a path that (combined with mutation and natural selection) brings about incompatible genetic makeup between the two formerly like models. That is, the two groups can no longer reproduce and create fertile offspring. Where one species existed, there are now two. Where there were two, there are soon four. Thus, one common ancestor is shared by foxes, wolves, bears, coyotes, and dogs. Similarly, *Eohippus* (the "dawn horse") branched off over a 50-million-year period into rhinos, tapirs, zebras, donkeys, and modern horses.

The same principles of natural selection apply to all organisms, from the smallest one-celled bacteria to the largest whale. For example, certain types of mites, which are microscopic spiderlike creatures, make their homes in the lungs of monkeys, noses of seals, and the breathing tubes of honey bees. Remarkably, mites of the species *Demodex folliculorum* live inside the pores of the skin cover-

Figure 13-2

A COMMON HARD TICK (FAMILY IXODIDAE), A TYPE OF MITE

ing our eyelids and faces. They eat, mate, and lay eggs in these minute, but obviously ideal, ecosystems.

There are thousands of different kinds of mites, and each has adapted to a specific niche in the ecosystem in order to survive. Having mites in your skin is not a sign of poor hygiene—it is an example of an organism's successful adaptation to an available niche. These mites evolved hundreds of millions of years before humans existed, as evidenced by certain species of mites that have been preserved in amber dating back to the Paleozoic Era (545–251 million years ago). Before humans evolved, *Demodex* lived in the skin pores of other animals such as dogs and sheep, and still live with those other species. The mouth parts of mites have adapted to pierce, suck, tear, and chew, depending on the niche in which the species is found. The evolutionary adaptation of mites is a story repeated for every species dispersed across this planet.

Each of those organisms has no "choice" but to develop the characteristics most necessary for survival. Some courses taken by certain species negatively impact other species, as we will see in Chapter 16, in the case of infectious bacteria. On the other hand, natural selection in the fungus kingdom led to one of the greatest discoveries in medicine. The antibiotic penicillin is made from the colorful mold that commonly grows on bread. This mold is a species of fungus *(Penicillium)* that adapted to life by producing molecules that kill many bacteria. The production of the substance essential for penicillin is simply the fungi's well-adapted defense against bacterial infection. This and numerous other examples of natural selection have greatly benefited mankind.

Darwin imagined the branches of the rich and widespread tree of life pulling itself inward as the evolutionary clock runs in reverse over millions of years. He speculated that all animals have "one common ancestor" and called his principle Natural Selection or Survival of the Fittest. "Natural Selection leads to the improvement of each creature in relation to its . . . conditions of life." Darwin correctly surmised that evolution is not random, but does rely on the powerful and deliberate influence of natural selection which, in turn, is determined by geology, climate, competition, and the other forces that comprise the total environment and that have been shaping forms of life since the dawn of time.

After Twenty Years in the Making, Darwin Publishes *The Origin of Species*

An Ape for a Grandfather

On January 29, 1839, Darwin married his cousin, Emma Wedgwood, and began a family that was to eventually number ten children, several of whom were to be distinguished in their own careers, including three who were knighted. Darwin supported the family by selling books and articles on biological subjects related to his collections and observations during the *H.M.S. Beagle* voyage.

During the years after the voyage, in a manner and outlook similar to the way Newton developed the basic laws of physics in the years after Woolsthorpe, Darwin secretly continued to compile his case for natural selection: "After five years' work I allowed myself to speculate on the subject, and drew up some short notes; these I enlarged in 1844 into a sketch of the conclusions . . ." He shared his theory with a few close friends, including geologist Charles Lyell, who were skeptical at first but eventually were persuaded. Because of Darwin's humble and cautious nature and Lyell's advice "never to get entangled in a controversy," Darwin's original plan had been to leave the manuscript upon his death with instructions to his wife on getting it published:

> I have just finished my sketch of my species theory . . . I therefore write this in case of my sudden death, as my most solemn and last request, which I am sure you will consider the same as if legally entered in my will, that you will devote four hundred pounds to its publication and . . . take trouble in promoting it. I wish that my sketch be given to some competent person, with this sum to induce him to take trouble in its improvement and enlargement . . . Mr. Lyell would be the best if he would undertake it; especially with the aide of Hooker . . .

Darwin had feared the inevitable uproar that would be created by such a theory. Thus, in Newton-like procrastination, he would not have publicly disclosed this revolutionary concept had it not been for the writings and similar views of the British naturalist Alfred Russel Wallace (1823–1913), which caused Darwin to come forth and explain:

> My work is now nearly finished; but as it will take me many more years to complete it . . . I have been urged to publish this Abstract. I have more especially been induced to do this, as Mr. Wallace . . . has arrived at almost exactly the same general conclusions that I have on the origin of species.

Prompted by Wallace's competing work, Darwin finished his monumental book, *The Origin of Species by Means of Natural Selec-*

tion or the Preservation of Favoured Races in the Struggle for Life. It was first published on November 24, 1859—more than 22 years after his first notes on the subject—and sold out immediately. It was re-published in revised editions, with the sixth and last edition ap-pearing in 1872. He gave credit to those who came before him and who set the stage for his own work. But it was Darwin's biological evidence that strongly suggested that existing species have evolved from earlier different forms, and it was Charles Darwin, not those he acknowledged, who figured out how natural selection produces adaptation, to the extent that it could be logically explained with-out knowing about genetics, DNA, or the detailed workings of cells.

From the moment of its initial publication in 1859, the book be-came the subject of intense and widespread controversy. For the next several years Darwin maintained a low profile while his friends and colleagues—Joseph Hooker, Thomas H. Huxley, Alfred Russel Wallace, and Charles Lyell, a group of scholars and professors who were respected and distinguished in their own right—manned the front line in the war against mythical beliefs.

Calling him "the most dangerous man in England," Darwin's critics and enemies comprised a diverse group. English geologist Adam Sedgwick (1785–1873), a professor at Cambridge and presi-dent of the prestigious Geological Society of London, never gave a sound reason for his opposition. He would only say that he rejected Darwin's use of hypotheses and deductive (as opposed to inductive) reasoning to prove his theories. Richard Owen (1804–1892), a British anatomist, paleontologist, Fellow of the Royal Society, and the nation's leading biologist at the time Darwin put forth his the-ory, was Darwin's most vehement and vicious detractor. A jealous and ambitious man, Owen unscrupulously attacked Darwin on nu-merous occasions. Many others who tried to discredit Darwin and his theory did so on the basis of their religious beliefs. If evolution were true, the account of the Creation in the Book of Genesis was false. To their further dismay, if natural selection worked automat-ically, there was no divine guidance in the life and growth of plants and animals.

These issues were debated at a meeting of the British Association for the Advancement of Science on June 30, 1860, in Oxford, with Owen's spokesman, Bishop Samuel Wilberforce (1805–1873), on one side, and Darwin's spokesman, Thomas H. Huxley (1825–1895), on the other. This confrontation symbolized the firm division between science and theology. In response to Wilberforce's patronizing comment during the debate, concerning Huxley's relation to apes and monkeys, Huxley gave his famous response:

> If . . . the question is put to me, would I rather have a miserable ape for a grandfather or a man highly endowed by nature and possessed of great means of influence, and yet who employs these faculties and that influence for the mere purpose of introducing ridicule into a grave scientific discussion—I unhesitatingly affirm my preference for the ape.

Amid Growing Controversy Darwin Applies Evolution to Humans

Slits on the Side of the Neck

Contrary to popular belief, and even though Huxley and Wilberforce debated human evolution just seven months after publication of *The Origin of Species*, that book does not address the evolution of humans. It's about plants and animals. In fact, there is just one brief comment by Darwin on *human* evolution, which appears toward the end of the book: "In the future I see open fields for far more important researches . . . Much light will be thrown on the origin of man and his history." Darwin's view on whether his theory also applies to humans was written between the lines, but he left it to other biologists and the public to raise the issue: "During many years I collected notes on the origin or descent of man, without any intention of publishing on the subject, but rather with the determination not to publish, as I thought that I should thus only add to the prejudices against my views."

In the mid-1800's, the belief that we are immutable and unchanging was as clear as the universally accepted "fact" in the mid-1500's that the Sun circles the Earth. To question the essence of our very being was an insult even greater than attacking the significance of our place in the universe. Yet that was the inevitable impact and result of *The Origin of Species*. Until Darwin, man was something special. Until Darwin, man was qualitatively different from animal. This Earth and this universe were created by God for man and no other reason . . . until Charles Darwin.

Twelve years after publication of *The Origin of Species*, controversy boiled anew. Darwin was compelled to pick up his pen once again to squarely address the subject of human evolution in *The Descent of Man*, published on February 24, 1871. Yet Darwin had been contemplating the issue at least since November 27, 1838, the date of his first notes applying natural selection to humans. In *The Descent of Man*, he explained how he'd abandoned his original intention not to become embroiled in the controversy: "Now the case wears a wholly different aspect," he wrote, citing the growing acceptance by respected naturalists of his basic thesis of natural selection. He felt the call "to put together my notes, so as to see how far the general conclusions arrived at in my former works were applicable to man."

The two-volume work totals nearly a thousand pages and is even more ambitious than *The Origin of Species*. "The sole object of this work is to consider, firstly, whether man, like every other species, is descended from some pre-existing form; secondly, the manner of his development; and thirdly, the value of the differences between the so-called races of man . . ."

Buoyed by the acceptance of his theory of evolution by natural selection, Darwin had grown bold in the twelve years that separated his two greatest books. In *The Descent of Man* he asserted that "man is descended from some less highly organized form . . . The great principle of evolution stands up clear and firm, when these groups of the facts are considered . . . such as the mutual affinities of the members of the same group, their geographical distribution in past and present times, and their geological succession." Regarding that geographical distribution, he predicted that the oldest fossil evidence of modern human's predecessors would someday be found in Africa:

> In each great region of the world, the living mammals are closely related to the evolved species of the same region. It is, therefore, prob-

able that Africa was formerly inhabited by extinct apes closely allied to the gorilla and chimpanzee; and as these two species are now man's nearest allies, it is somewhat more probable that our early progenitors lived on the African continent than elsewhere.

Because of the lack of fossil evidence during Darwin's lifetime and because the "Dark Continent" was considered unfit as a place of man's origin, this prediction was considered folly all the way into the mid-20th century. Yet his fundamental theory was soon widely accepted:

> Man is developed from an ovule, about the 125th of an inch in diameter, which differs in no respect from the ovules of other animals . . . The embryo itself at a very early period can hardly be distinguished from other members of the vertebrate kingdom . . . [T]he slits on the side of the neck still remain . . . and the os coccyx projects like a true tail extending considerably beyond the rudimentary legs . . . We thus learn that man is descended from a hairy quadruped, furnished with a tail and pointed ears, probably arboreal in its habits . . . probably derived from an ancient marsupial animal, and this through a long line of diversified forms, either from some reptile-like or some amphibian-like creature, and this again from some fish-like animal . . .
>
> Man may . . . feel . . . some pride at having risen . . . to the very summit of the organic scale . . . But we are . . . concerned . . . only with the truth as far as our reason allows us to discover it. I have given the evidence to the best of my ability; and we must acknowledge . . . that man with all his noble qualities, with sympathy which feels for the most debased . . . with his god-like intellect which has penetrated into the movements and constitution of the solar system—with all these exalted powers—Man still bears in his bodily frame the indelible stamp of his lowly origin.

Fossil Records Confirm Evolution

A Fondness for Beetles

Darwin created the foundation that biologists, paleontologists, and geologists later built upon to explain the evolution of all life. The science of geology established "eras" primarily to reflect major changes in the development of life on Earth by designating geologic time spans of great magnitude. The eras are divided into periods, which are further broken down into epochs. We are presently in the Holocene Epoch (which began 10,000 years ago) of the Quaternary Period (which goes back 2.5 million years) of the Cenozoic Era (which covers the last 65 million years). Here is a summary of the evolutionary history of all life as revealed by the fossil evidence, divided into each of the four geologic eras.

Precambrian Era (4.6 billion to 545 million years ago): Life emerges from the primordial soup about four billion years ago, in the form of a bacterial cell. For the next three billion years, bacteria cells colonize Earth and remain the only life. About 600 million years ago, the first multicellular life evolves.

Paleozoic Era (545 to 251 million years ago): An incredible explosion of life begins 545 million years ago with the diversification of underwater life, leading to shelled animals 425 million years ago, followed by land animals 395 million years ago, the evolution of amphibians from fish about 350 million years ago, and reptiles beginning 280 million years ago.

Mesozoic Era (251 to 65 million years ago): After the most devastating extinction in history, 250 million years ago, the dinosaurs evolve and dominate the Earth from 230 to 65 million years ago, then die out.

Cenozoic Era (65 million years ago to present): After another mass extinction, the Earth's climate cools and vast rain forests give way to grasslands and woods. Mammals diversify into numerous forms. About four million years ago, the first members of the human family of primates emerge in Africa.

Starting about 545 million years ago (during the Cambrian Period of the Paleozoic Era), evolution experienced its own "Big Bang"—there was an explosive radiation of life in which the predecessors of nearly all the species that now blanket Earth (and others that are now extinct) first appeared. The famous biologist Stephen Jay Gould wrote about the Cambrian explosion in his book *Won-*

derful Life: The Burgess Shale and the Nature of History, which explores the pace and breadth of evolution. Until recently, paleontologists and biologists believed that the Cambrian explosion of life lasted anywhere from 20 million to 70 million years. However, in 1993, a team of scientists from Harvard University, the Massachusetts Institute of Technology, and the Yakutian Geoscience Institute in Yakutsk, Russia, found evidence in Cambrian rocks indicating that the period of rapid diversification lasted a much shorter time, perhaps only five million years or less.

This great diversity of plant and animal life forms flourished for over 200 million years, and then suffered a massive setback about 250 million years ago when *90 percent* of all ocean life and a large percentage of land species were destroyed—the most massive and devastating extinction ever recorded and the beginning of the process that allowed for the ascension of the dinosaurs. The fossil record reveals that in some areas as much as 97 percent of all trees had died by the end of the Permian Period (250 million years ago), resulting in the proliferation of a wood-rotting fungi that virtually covered the planet. Some paleobotanists and other scientists believe that a series of volcanic eruptions in Siberia caused this extinction. The sediments show that over a one-million-year period, one volcano in particular poured out the equivalent of so much molten basalt that it could have paved the entire Earth with a layer six yards deep. The carbon dioxide and acids released from such eruptions could have poisoned the air and water, thus causing the extinctions.

In 1996, four researchers proposed a different theory—that a huge upwelling of deep ocean water laden with carbon dioxide caused the Permian catastrophe. These biologists and geochemists maintain that this dangerously high level of carbon dioxide was natural at the ocean depths but entered the surface sea water and the atmosphere over a period of many years because of a "global turnover of deep ocean water," as now reflected in the pattern of calcium carbonate sediments laid down at the end of the Permian Period. These excessive amounts of carbon dioxide would have caused carbolic acid to form in animals' blood in lethal doses and could have produced a greenhouse effect across Earth, destroying marine animals in massive numbers and upsetting many other species' delicately balanced environments.

The mass extinction at the end of the Permian Period gave rise to the dominance of the dinosaurs for about the next 140 million years. The dinosaurs were wiped out by yet another extinction (at the end of the Cretaceous Period), most likely caused by a massive asteroid or comet that landed near the Yucatan Peninsula, resulting in a shroud of dust that encircled the globe and wreaked havoc on the climate and vegetation on which the dinosaurs relied, restricting their habitat and squeezing them to extinction. But this extinction opened up new ecological niches. With the forces of natural selection at work, life rebounded with greater success and diversity.

The latest burst of life, following the extinction of the dinosaurs, surrounds us. It is seen by examining a pinch of soil or by looking at the vast number of species that now inhabit Earth. With the aid of a microscope, you would find about ten billion of the longest and most enduring residents of planet Earth: *bacteria*. These one-celled organisms (representing hundreds of species in those ten billion bacteria) break down the organic material in the soil. Overall, biologists have identified and named over 1.4 million species of animals, plants, and microorganisms, although most experts estimate that literally millions of species of bacteria and insects, and even some fish, birds, and mammals have not yet been discovered. Biologists expanded Carl Linnaeus's original two-kingdom scheme (plants and animals) into today's five-kingdom system, with fungi (Latin for "mushrooms"), protozoans (meaning "first animals"), and bacteria earning their own kingdoms because they cannot be considered plant or animal by virtue of their cell structure or inability to photosynthesize.

The following list shows the large number of insect species in comparison with all other life forms, including the extraordinary number of beetle species. When asked what the study of creation had revealed about God's plan, Darwin's friend Thomas H. Huxley quipped, "That the Almighty has an inordinate fondness for beetles."

- 751,000 insect species (of which 290,000 are beetles)
- 281,000 total species of amphibians, reptiles, fish, birds, and mammals

- 275,000 plant species
- 69,000 fungi species
- 31,000 protozoan species
- 5,000 bacteria species

At the end of *The Origin of Species*, Darwin wrote, "There is a grandeur in this view of life . . . Whilst this planet has gone cycling on according to the fixed law of gravity, from so simple a beginning endless forms most beautiful and most wonderful have been and are being evolved."

In *The Origin*, Irving Stone wrote that at the end of Darwin's life:

> Perhaps it was the long bouts of illness; perhaps the concentrated periods of incredibly hard work; perhaps it was his snow-white, all-encompassing beard . . . but Charles Darwin now looked as though he were the Patriarch of the World. His eyes were sunken; he carried the look of a man who was about to say farewell; and did not mind at all.

Throughout his life Darwin had suffered from a number of physical ailments, including episodes of depression, and a heart attack in 1873. On April 19, 1882, he suffered another heart attack and died. Upon petition by twenty members of Parliament, Charles Robert Darwin was laid to rest in Westminster Abbey. Among the pallbearers were Hooker and Huxley.

He had come a long way from the poor student whose father feared would be "a disgrace to the family," and from the collector of specimens on the *Beagle*. He evolved from the Great Describer to the Great Interpreter. He kept asking until there was an explanation that made sense. He collected the unordered facts available and made them coherent and understandable. In his own lifetime he not only authored evolutionary theory, but he succeeded in changing the public's opinion of it from an outrageous heresy to a generally accepted principle. Entire scientific disciplines have grown directly out of Darwin's work. The field of genetics (discussed in Part Six) and the search for the ultimate mechanism of heredity and evolution—the DNA molecule (Part Seven)—have now combined with Darwin's work to form a complete picture of evolution, resulting in a fundamental change in our concept of all life, including the human race.

Chapter Fourteen

Shaper of the Landscape

"Every animal leaves traces of what it was. Man alone leaves traces of what he created . . . Man is not a figure in the landscape. He is a shaper of the landscape.

How did the hominids come to be . . . dexterous, observant, thoughtful, passionate, able to manipulate in the mind the symbols of language and mathematics both, the visions of art and geometry and poetry and science? How did the ascent of man take him from those animal beginnings to that rising enquiry into the workings of nature, that rage for knowledge . . . ?"

JACOB BRONOWSKI
The Ascent of Man (1973)

Charles Darwin formulated the theory of evolution by natural selection. But in his day no one had any idea how long it had taken for humans to evolve, or through what stages it had occurred. Darwin opened the book on the evolution of humans. Others continue to write the story. The chapters that anthropologists and paleoanthropologists have written in this century represent yet another turning point in the history of human knowledge, and in our concept of our primitive ancestors.

Diverse Fields of Science Combine in the Twentieth Century to Define the Course of Human Evolution

Mammals Thrive on December 30

Paleoanthropology is the study of early humans, including fossils of hominids (humanlike beings) and extinct primates from which we evolved, as well as relics of the cultural activities of ancient humans. As we learned in Chapter 13, dinosaurs became extinct about 65 million years ago when the Earth's climate cooled and mammals began to flourish across the Earth. With the wealth of physical information discovered in recent decades, together with the application of unprecedented technology, 20th-century paleoanthropologists have connected the human and nonhuman links of this 65-million-year-old chain by combining developments in several disciplines.

- *Physics and geology:* Through 20th-century developments in physics (radioactive dating) and geology (geological stratification), we are now able to measure the age of fossils with great accuracy.
- *Biochemistry and biology:* In Parts Six and Seven we will see how biochemists and biologists discovered how the fundamental "code" of reproduction was passed from primitive life to humans.
- *Astrophysics and geochemistry:* Even astrophysics and geochemistry have helped put pieces of the evolutionary puzzle together through the study of the formation of elements in space and the study of the chemical composition of the early Earth.

Thus, we've been able to recreate the evolutionary process of humans through a remarkable synthesis of diverse disciplines that were once thought to be unrelated. As Darwin predicted, the process began with "one common ancestor" well before the appearance of reptiles and amphibians. This evolutionary path has been ongoing for billions of years. To help conceptualize the history of evolution, we will squeeze the 15-billion-year history of the universe into an imaginary year, with the Big Bang occurring at the first instant on the morning of January 1 and the year 2000 beginning at the first instant after December 31. These would be the significant dates of that "Year of the Universe" *before* the month of December:

Table 14-1

YEAR OF THE UNIVERSE—THROUGH NOVEMBER

January through November (13.75 billion years)

> January 1 — Big Bang
> May 1 — Origin of Milky Way Galaxy
> September 9 — Origin of Solar System
> September 14 — Formation of Earth
> September 25 — First Life on Earth
> October 9 — Oldest Fossils (bacteria and algae)
> November 12 — Oldest Fossil of Photosynthetic Plant
> November 15 — First Cells with Nuclei

Since each month of that year represents about 1.25 billion years, the first life doesn't appear until September 25, and *recorded history doesn't begin until the last minute before midnight on December 31.* The other significant events that occurred during the 1.25 billion years of the month of December are shown in Figure 14-2.

Sixty-five million years ago on this calendar is December 30. This is very recent, compared to the age of the Earth at 4.6 billion years and the 15-billion-year-old universe. Thus, just "yesterday" the dinosaurs disappeared and human predecessors began to emerge. The hominid's tenure on Earth—a few million years—is substantially more brief than the 140-million-year reign of the dinosaurs. *In our modern form we've been around only about 40,000 years.*

Our Ancestors Are Revealed in the Fossil Records

Rodents in the Jungle of Europe

Sixty-five million years ago the world was covered with tropical forests from the equator to latitudes as high as what is now north-

Figure 14-2

December in Year of the Universe

December

Sunday	Monday	Tuesday	Wednesday	Thursday	Friday	Saturday
	1 Oxygen atmosphere begins to develop on Earth.	2	3	4	5	6
7	8	9	10	11	12	13
14	15	16 First worms.	17 (Paleozoic Era and Cambrian Period begin.) Invertebrates flourish.	18 First oceanic plankton. Trilobites flourish	19 (Ordovician Period) First fish and first vertebrates.	20 (Silurian Period) First vascular plants. Plants begin colonization of land.
21 (Devonian Period begins.) First insects. Animals begin colonization of land.	22 First amphibians and first winged insects.	23 (Carboniferous Period) First trees and first reptiles	24 (Permian Period begins.) First dinosaurs.	25 (Paleozoic Era ends and Mesozoic Era begins.)	26 (Triassic Period) First mammals.	27 (Jurassic Period) First birds.
28 (Cretaceous Period) First flowers. Dinosaurs become extinct.	29 (Mesozoic Era ends/Cenozoic Era and Tertiary Period begin.) First cetaceans and first primates.	30 Early evolution of frontal lobes in the brains of primates. First hominids. Giant mammals flourish.	31 (End of the Pliocene period. Quaternary Period begins.) First humans.			

ern France and Germany and southern Alaska. A wide variety of plant and animal life thrived, including a rapid proliferation and evolution of mammals. Among them were the predecessors of the primates, including what are now called ancestral primates. They do not resemble their successors—the monkeys, the apes, the human beings—but over a period of 65 million years, it takes very small modifications by natural selection to turn a long-tailed ro-dentlike mammal (yes, your ancestor) into a totally different form of life. The closest relatives to our most ancient mammalian ances-tor that still exist today are called prosimians, another type of pri-mate. Prosimians include lemurs, lorises, and tarsiers, all of which live in the jungle. Fossils have shown that certain forms of prosimi-ans have changed very little over time, while other ancestral pri-mates branched off and obviously went through radical changes.

Tracing human evolution from the time of that little, scuttling mammal is like watching an incredible slow-motion film. We can examine each stage of that 65-million-year-long story:

- Over the first few million years of the *Paleocene Epoch* (which lasted from 65 million to 54 million years ago), it evolves into a tree-dweller and eats fruit and insects.
- Individuals with longer hind legs are naturally selected as the species adapts to a climbing, swinging, arboreal life.
- Flat nails replace ratlike claws, and it begins to develop a thumb that can be opposed to the fingers.
- The fingers grow longer, more flexible and able to grip.
- During the *Eocene Epoch* (54 million to 38 million years ago), the posture becomes more upright, the head can now rotate to a greater extent, and the brain grows. (See Figure 14-3.)
- Eyesight improves and eyes grow larger and widely spaced, while the sense of smell decreases—indicating natural selec-tion against smell and in favor of vision; accordingly, the snout becomes shorter.
- During the *Oligocene Epoch* (38 million to 26 million years ago), a branch of the prosimians begins to look much like monkeys or apes. They live in the tropical bush country of what is now Egypt.

Figure 14-3

PALEOCENE AND EOCENE ANCESTRAL PRIMATES

Loris found in Africa and Asia—Similar to Paleocene ancestral primate of 65 million years ago

Skull drawn from fossils of Eocene ancestral primate (top) of about 45 million years ago, and restoration of the head (bottom)

- The teeth evolve into a four-cusp pattern unique to primates.
- The frontal bone above the eyes and around the eye sockets approaches orbital closure, similar to that of humans.
- During the early part of the *Miocene Epoch* (26 million to 7 million years ago), the creature grows to about a foot tall, and continues to increase in size as an adaptation to avoid predators.
- It evolves out of the trees and returns to the ground because fruit is less abundant. The diet changes to berries, roots, and insects. The animal begins to walk on the soles of its feet and knuckles of its hands. (See Figure 14-4.)
- Skeletal adaptations occur consistent with the new mode of moving, eating, and living.
- The jaw becomes smaller and the face more vertical.
- By the end of the late *Miocene Epoch* (about 7 million years ago), the animal walks more upright, and is now about three or four feet tall—a small ape, to be known as *Ramapithecus*.

Figure 14-4

OLIGOCENE AND MIOCENE ANCESTRAL PRIMATES

Aegyptopithecus flourished during the Oligocene Epoch, 26 to 38 million years ago. This reconstruction is based on numerous fossils of this 12-pound creature found in the twentieth century.

Proconsul lived during the early Miocene Epoch, specifically between 14 and 23 million years ago.

- During the *Pliocene Epoch* (7 million to 2.5 million years ago), the being diverges from African apes and evolves into ape-man, as it develops a larger brain and several other hominid (meaning humanlike) characteristics, particularly the ability to walk on two legs. Twentieth-century scientists will name it *Australopithecus*. It is hairy and still spends some time in the trees, but it moves in bands, hunts with clubs, and is agile, keen-sighted, alert, and more intelligent than the baboons and apes with which it competes for food in the African savanna. It does not have a language, but communicates with an array of sounds, gestures, and nonverbal expressions. Its life span is 15–20 years. (See Figure 14-5.)

Figure 14-5

RAMAPITHECUS AND *AUSTRALOPITHECUS*

Ramapithecus,
late Miocene Epoch,
seven million years ago

Australopithecus family
four million years ago

- Earth then enters the *Pleistocene Epoch* (beginning 2.5 million years ago) of the present Quaternary Period. Over the next million years, the creature becomes less apelike and more humanlike. Someday, it will be called *Homo habilis*. It begins to make crude stone tools and to construct shelters, indicating further biological changes in the hand and the brain centers that control the hand. About 100,000 individuals exist at any given time during the period when *Homo habilis* is thriving.
- Over the course of the next several hundred thousand years, the animal begins to stand tall (about five feet). Its brain is now twice the size of *Australopithecus*'s brain. During the period beginning 1.5 million years ago and extending to 400,000 years ago, this new species, *Homo erectus*, spans the transition between *Homo habilis* and early *Homo sapiens*. It is the first hominid to leave Africa, and it begins to populate the rest of the

Earth. It disperses across several continents, and grows to a population of about one million at any given time.

In the 20th century, *Homo erectus* fossils will be found in China, Java, Africa, and Europe, all showing a cranial capacity of about 60 cubic inches, a flat, receding forehead, large brow ridges, a flattened skull vault, large teeth and jaws, and a small chin. It builds huts, and the more recent ones (beginning 700,000 years ago) build fires and cook their food. They invent tools for chopping wood, cutting meat, skinning animals, and scraping hides. (See Figure 14-6.)

Figure 14-6

HOMO HABILIS AND *HOMO ERECTUS*

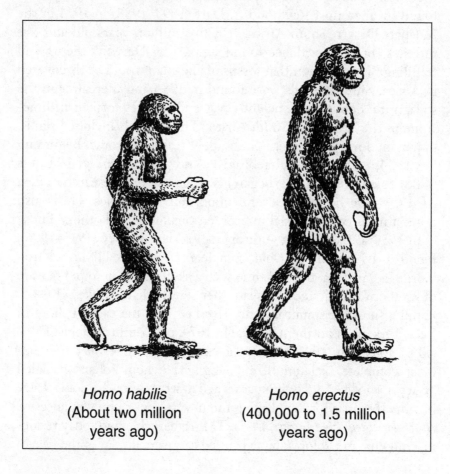

Homo habilis
(About two million
years ago)

Homo erectus
(400,000 to 1.5 million
years ago)

Homo erectus speaks in simple sentences and develops a more complicated social organization than that of its ancestors. Labor is divided between the sexes, the family unit takes on significance, bands of 20 to 50 individuals live together, and permanent home sites are sometimes established. Its life span is 20–30 years, not much less than its future relatives of 14th-century Europe, whose average age at death will be 38.

The fossil record of *Homo sapiens'* ancestors is not completely clear. There are gaps and questions, which are filled by assumptions and interpolations. The chronology described above is primarily based on fossils discovered over the last seventy-five years, such as the Taung baby skull (named after the Taung limestone quarry) found by anatomist Raymond A. Dart (1893–1988) in 1924 in the Kalahari Desert, South Africa. It is two million years old and was the first concrete evidence found supporting Darwin's theory and fulfilling his prediction that fossils of ape-man would be discovered in Africa. Similar fossils were found in 1936 and thereafter in the stalagmitic deposits of ancient caves and fissures formed in limestone in the Transvaal, South Africa. These fossils included skulls, upper and lower jaws, teeth, limbs, pelvis, and vertebrae. Beginning in 1924, Kenyan archaeologist and anthropologist Louis S. B. Leakey (1903–1972) began his archaeological research in East Africa, where he made several important discoveries. He is also noted for his controversial interpretations of his and others' finds.

In 1959, Leakey's wife, anthropologist Mary Leakey (1913–1996), found a 1.7-million-year-old complete hominid skull in Olduvai Gorge in Tanzania, the first time such a skull had been found outside of southern Africa. Ten years later her son, Richard Leakey (1944–), found a similar cranium in a dry riverbed near the eastern shore of Lake Turkana in northern Kenya. In 1974, paleoanthropologist Donald C. Johanson (1943–) made the extraordinary discovery of an almost complete skeleton of a young girl (whom Johanson called "Lucy") who lived 3.5 million years ago in what is now Ethiopia. Lucy became the basis for establishing the new species of human ancestor *Australopithecus* (See Figure 14-5). This remarkable find today resides in a museum in Ethiopia, but a painted cast is on display at the Cleveland Museum of Natural History, where Johanson was working as a curator of anthropology at the time of his discovery. During the period between 1988 and 1995, anthropologist Meave Leakey (1942–) (Richard's wife and co-researcher) made significant fossil discoveries, including teeth, jawbones, and parts of legs of four-million-year-old hominids at Lake Turkana, illuminating our knowledge of bipedalism and brain size of our ancestors.

There have been numerous other discoveries in Africa since 1924—particularly from the Omo region of southern Ethiopia and from the Lake Turkana area. The gaps in the story continue to narrow, regarding when and how our line of primates branched off, destined to be the most intelligent living thing known in this corner of the universe.

These early ancestors who wandered the plains on two legs were physically suited for a nomadic life. Those who created the most successful hunting strategies and found the safest havens for the family were the most likely to survive. The strongest, fastest, and, foremost, the most intelligent members of the tribe or band were more likely to pass their genes on. Just as the fastest and fiercest lion prevails among lion species, the most intelligent hominids and humans prevailed among our ancestors and developed into modern *Homo sapiens.*

Table 14-7

DECEMBER 31 OF YEAR OF THE UNIVERSE

Origin of prehistoric ancestors of humans	1:30 p.m.
First humans	10:30 p.m.
Widespread use of stone tools	11:00 p.m.
Beginning of last glacial period	11:56 p.m.
Invention of agriculture	11:59:20 p.m.
First cities	11:59:35 p.m.
First dynasties in Egypt	11:59:35 p.m.
Development of astronomy	11:59:35 p.m.
First writing	11:59:51 p.m.
Birth of Christ	11:59:56 p.m.
Fall of Rome	11:59:56 p.m.
Renaissance in Europe	11:59:59 p.m.
The 21st Century	The First Instant of New Year's Day of the Year 2000

Returning to the Year of the Universe model again, the preceding timetable identifies the significant events of the afternoon and evening of December 31. We'll now examine paleoanthropology's finds that have given us a clear picture of the first humans who evolved beginning at about 10:30 p.m. on this last day of the Year of the Universe.

Neanderthal Man Emerges, Then Disappears

Rickets, Arthritis, and the Newcomers

In 1856, in a cave near Düsseldorf, Germany, a science teacher by the name of Johann Carl Fuhlrott found fourteen pieces of bone that looked like those of a distorted human. The skull was long, low and wide. It suggested that this being's face had heavy brow ridges, a receding forehead, small cheekbones, and large teeth. The chest was broad, and the limbs were heavy, with curving thigh and forearm bones, larger-than-human feet and hands, and short fingers and toes.

This was the first ancient human fossil skull ever to be discovered. It was named Neanderthal, after the Neander River Valley, Germany, where the bones were found. The first edition of *The Origin of Species* appeared three years later, in 1859. But even then, no one was prepared to accept what appeared to be a primitive animal as his or her ancestor, or to accept the idea that early humans lived in caves. Fuhlrott himself concluded that the bones were probably distorted from the weather and that they belonged to some unfortunate person who was washed into the cave by Noah's flood. Other scientists who examined the skull specimen concluded that it was a very ancient pathological idiot. The highly respected German anatomist, pathologist, and statesman Rudolph Virchow (1821–1902) wrote that the bones were from a Mongolian cavalryman who suffered from rickets in childhood and arthritis in old age, and who had been clubbed on the head. Even Darwin's friend and supporter, Thomas H. Huxley, the world-renowned biologist, studied the find and said, "In no sense can the Neanderthal bones be regarded as the remains of a human being intermediate between men and apes." This effectively silenced all speculation for the time being.

In 1886 two more Neanderthal skeletons were found, this time near Spy, Belgium. Primitive stone tools and remains of extinct animals were discovered alongside the skeletons. Though Virchow repeated his theory about diseased specimens, many scientists were beginning to recognize that the idea of primitive ancestors was valid although distasteful. More complete Neanderthal fossils were found in France, beginning in 1908. In subsequent years, remains

Figure 14-8

NEANDERTHAL

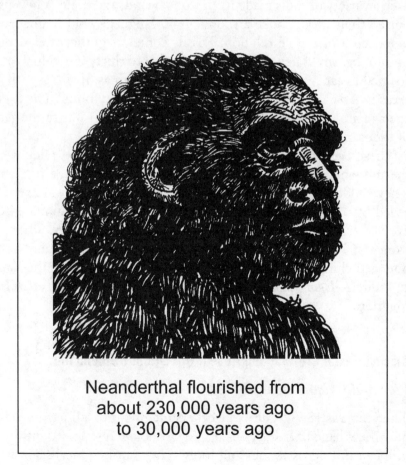

Neanderthal flourished from
about 230,000 years ago
to 30,000 years ago

were found throughout Europe, Asia, and Africa, and now total about three hundred individuals.

Today, most paleoanthropologists agree that the Neanderthals were the ancestors of *Homo sapiens* from the period beginning about 230,000 years ago until they vanished 30,000 years ago. Neanderthal bones have played a key role in studying the evolution of humans. These people seem to have been nearly as intelligent as modern man. They had a three-pound brain, which is as large as the brain of modern humans, and stood as erect as *Homo sapiens*. Surviving in the harsh and frigid wilderness of intermittent Ice Ages, they were cave-dwellers; they used fire, including torches and lamps; they hunted with spears and occasionally caught fish; they used a variety of tools made from stone, wood, and bone. They also buried their dead, spoke a rudimentary language, and might have practiced a primitive religion. Fossils contain significant evidence of cannibalism. They lived in groups of up to thirty individuals and roamed as far north as Britain and as far south as Spain, as well as areas in Asia and Africa where remains have been found. The average age at death was 30, and their total population was in the tens of thousands at the height of their existence.

These early modern humans followed *Homo erectus*. They were the individuals generally regarded to be the Stone Age humans. They suddenly disappeared from the fossil record and were superseded by a new, more advanced culture that either annihilated or evolved from the Neanderthals. The "newcomers" were different from the Neanderthals. They had a high forehead, a more pronounced chin, and a flatter face. Anatomically, they were the same as modern *Homo sapiens* and someday would be known as Cro-Magnon.

Discoveries of Cro-Magnon Reveal Our Direct Descendants

Clay and Ground Bone

The year was 1868. Railroad workers digging into a limestone cliff near the village of Les Eyzies in the Dordogne region of southwest France came across bones and tools. They notified local scientists, who uncovered four human skeletons: a middle-aged man, a younger man, a young woman, and an infant. They also found flint tools and weapons, and ornaments made of shells and animal teeth. The cave was named Cro-Magnon, after a local hermit named Magnon who had lived there. The term now generally refers to the modern humans who lived at various places around the Earth 10,000 to 40,000 years ago. Most paleoanthropologists believe that the Neanderthals evolved into Cro-Magnon, while others believe that the shorter, heavy-boned Neanderthals comprised a specialized "side branch" of the human line that went from *Homo erectus* to Cro-Magnon to *Homo sapiens*. The paleoanthropologists who subscribe to that theory believe that the Neanderthals became extinct because they could not compete during the 10,000 years during which they overlapped with Cro-Magnon (30,000 to 40,000 years ago), and that Cro-Magnon is the *only* living direct link to *Homo erectus*. Arguments for both theories are based on differing interpretations of the many Neanderthal and Cro-Magnon fossils, tools, and weapons found throughout the world, which date back to these periods. Under either theory, we are clearly Cro-Magnon's direct descendants.

Though many of the Cro-Magnon people lived in caves, others built shelters. They developed fishing and hunting tools, such as nets, spears, and traps. They had impressive artistic ability, as seen in their cave art and sculptures made of stone, bone, and ivory. They sewed clothes, made footwear, and knew how to make fire. They had complicated spiritual ceremonies and rites for hunting, burials, and other customs and activities. At the site of Dolni Vestonice in the Czech Republic, they built kilns as long as 27,000 years ago, to fire a mixture of clay and ground bone into ceramics— mankind's first example of combining and treating two dissimilar substances to make a new and different product. It was not until 15,000 years later in Japan that any other group learned to turn clay into pots. The lives and lifestyles of Cro-Magnon were very similar to those of the North American Indian, the Australian Aborigine and a number of African tribes. At any one time, the Cro-Magnon people numbered about three million. They set the stage for our emergence into modern civilization, as we entered the last minute of the Year of the Universe.

It took almost four million years for humans to evolve from the little dark creature with the club in its hand—that is, *Australopithecus* in central Africa—to the modern form of *Homo sapiens*. But it is only in the last 30,000 to 40,000 years that our species has had the physical and intellectual attributes that we now regard as those of human beings. Cro-Magnon men and women of 30,000 or 40,000 years ago were as intelligent as *Homo sapiens* of today. If you could pluck an infant from a Cro-Magnon community of 30,000 years ago and raise him or her in your home, it is likely that the child would be no different in life and school from a modern child. This would not be true of Lucy or a Neanderthal child. We have simply built on the knowledge and experience of each generation to achieve the technological advances and create the social institutions that characterize modern civilization.

Creation of Distinct Races Is a Result of Natural Selection

Homo erectus *and Humidity*

Unique physical traits develop *within* each species as a result of surrounding environmental pressures, in a process similar to reproductive isolation, which was discussed in Chapter 13. Millions of years after hominids had become reproductively isolated from other primates, differences among them began to emerge because of their geographic isolation. They adapted to those unique environmental circumstances as the species gradually changed into *Homo sapiens*. In the most likely scenario, small groups or bands of *Homo erectus* migrated from Africa to various places all around the globe, then lived in isolation from one another as they evolved toward modern man, fighting to survive in a hostile world. However, that period of isolation was not long enough to result in speciation—the creation of separate species. That is, humans of each major race can mate with individuals in any other race and produce fertile offspring.

With the knowledge we now have about the gradually shifting

Figure 14-9

PROBABLE ROUTES OF EXPANSION OF *HOMO ERECTUS*

Approximate extent of
polar ice sheet during period

Lake Turkana
Olduvai

Java

crust of the Earth, the cycles of glaciation, and the effect of the environment on each species' evolution, we can understand how the geographically isolated groups of *Homo erectus* began to develop unique characteristics. For example, since overexposure to ultraviolet rays is harmful to skin, people who lived nearer to the equator formed a darker skin pigment than other people. The 19th-century German biologist and physiologist Carl Bergmann (1814–1865) proved that people in colder regions are bulkier than those in warmer areas because the icy wind caused their bodies to evolve as shorter and thicker to better preserve heat. Also, that environment caused their eyes to become squinted and protected by thicker eyelids. Bergmann's studies led to the principle known as Bergmann's Rule, which correlates external temperature and the ratio of body surface to weight in warm-blooded animals.

Other traits also represent valuable adaptations to the environment. For example, nose shape is related to humidity, because moistening the air is a prime function of the nose. If the moisture content of the air is not brought to about 95 percent relative humidity as the person inhales, the lungs can be damaged. People in dry areas (such as Asians) tend to have narrow noses, the shape that causes the body to add more humidity while inhaling. Central Africans, on the other hand, have broad noses. A myriad of other physiological traits—such as height, hair color and texture, heart rate, bone growth, fingerprints, and blood type—have also been studied and are attributable to environmental factors to which our ancestors adapted over a period of about a million years.

Modern scientists have been able to "watch" the process of natural selection in action in animals. For example, in November 1994, a biologist at the University of British Columbia reported on a population of stickleback fish that began to change shape and feeding habits in just a few generations when a new competitor forced the fish toward a different ecological niche. Countless studies of other animals, including 20th-century research on the finches of the Galapagos Islands, have provided objective observable data on this slow and silent process.

Geographical and climatic factors affect human evolution, as well as other external conditions, such as predation, disease, migration, conflict, and competition for breeding space, living space, and food. Over the million-year existence of *Homo erectus*, in response to all these stimuli, the geographically isolated groups developed distinct gene pools that physical anthropologists identify today as the nine major geographic races: (1) European Caucasoid; (2) African Negroid; (3) Asian Mongoloid; (4) Australoid; (5) American Indian; (6) Asian Indian; (7) Polynesian; (8) Melanesian; and (9) Micronesian. At the same time, each group developed its own language, culture and history.

The Existence of Distinct Races Has Had Major Political Effects

The Door of the Gods

The simultaneous physiological and cultural development and differentiation of the various racial groups has now revealed itself as one of the most significant *political* facts in human history. About 10,000 years ago, man changed from forager and hunter to animal raiser and farmer. Foragers and hunters traveled in bands averaging forty individuals, following herds of animals to ensure a food supply. But after the last Ice Age ended, there was a burst of new vegetation, and humans began to remain in one spot to domesticate animals and cultivate plants. We became shapers of the landscape. This was a significant change in the cultural and behavioral evolution of our species, because it was then possible for people to create villages and communities and, indeed, create civilization as we conceive of it today. Prior to that change, it was difficult for technology to develop to any significant extent because the nomadic people had to carry everything with them on their daily journeys following the herd. In a culture limited to simple devices that had to be light, accessible every night, and packed away every morning, there was no room or time for creativity, innovation, experimentation, or expansion of the mind. Until man settled down and created stable communities, the focus of daily life was survival. His only ambitions were to follow the traditions of past generations and create future generations.

Until about 10,000 years ago, mankind's only inventions were the basic tools for hunting and fishing, and perhaps the kiln of Cro-Magnon. At about this time, copper smelting was developed, and the wheel was invented in the Middle East or in the southern part of the former Soviet Union. Of course, the wheel contributed greatly to the onrush of technological development that began a few thousand years later, and eventually led to the use of rollers and pulleys in farming and for many other purposes. The invention of the wheel was followed by techniques for tanning hides, weaving, and making pottery. About 6,000 years ago, we learned how to combine copper and tin into bronze. Iron smelting was developed about 3,500 years ago. By about the year 1000 B.C., steel was being made in India, and its use spread to other parts of the world. The advent of permanent settlements and new technology had an extraordinary effect on world population, which increased from about five million to eighty-six million. This was more than a seventeen-fold increase in a period of 4,000 years—from 10,000 to 6,000 years ago.

By about 3000 B.C., man had trained the ox and horse to draw carts, and by 2000 B.C., we discovered how to ride the horse and sail the seas. With boats, horses, wagons, roads, and bridges, and with an expanding population and an adventurous spirit, people began to cross mountain ranges, travel down rivers and into seas. They began to traverse the distances that separated these once isolated groups. As nature dictates, these groups competed. They fought over territory and food, and for less important reasons. People spread out in cities and villages across the land, resulting in increasing contact with people who spoke a different language, who were lighter or darker, taller or shorter, believed in a different god, prayed to a different idol, dressed oddly, and, because of such differences, were often judged to be less human. As competition increased, and territorialism evolved into nationalism, those differences have been used to justify oppression and war.

In the European provincial view of the world depicted in the literature of the Middle Ages, Europeans regarded physical deviations from the European "norm" as the result of degeneration or divine punishment for sin. The history of most other human cultures is replete with similar narrow-mindedness and self-aggrandizement. Cultures and tribes all over the Earth called themselves "the people" or "all men," and relegated other groups of humans to subhuman status. Mediterranean means the "middle of the Earth." For thousands of years China called itself the Middle Kingdom. In ancient Greece, Mt. Olympus (abode of the gods) was the center of Greece, which was the center of the universe. The Hindus placed Mt. Meru in the center of the universe, much like Earth was believed to be the center of the universe until Copernicanism finally took hold in the Renaissance. Babylon (meaning "door of the gods") represented the location where the gods came down to Earth. For the Moslems, their Ka'bah was the highest point on Earth.

History, of course, is a complex subject. Conflicts throughout history result from a great variety of facts and circumstances at a given point in time. However, the physical and cultural differences among groups and nations undoubtedly has been one of the primary bases for such conflicts over the last several thousand years. This includes the conflicts among racial groups within the United States and other countries. The conflicts began with the earliest tribal battles and swept into the 20th century with incredibly brutal force. Religious and cultural differences have been the root cause of more deaths in the 20th century than in all prior centuries combined.

Because *Homo erectus* dispersed, then evolved into multiple races and cultures, world history overflows with accounts of military conflicts, the domination and enslavement of peoples throughout the world, and outspoken hatred between physically and culturally diverse groups. If we had understood before now the real mechanisms and history of human evolution, would there have been less death and destruction? If people understood that their appearances and cultural differences are the result of evolution, and are not divinely inspired or a reflection of innate superiority, perhaps they would not be so quick to condemn or so eager to torture, maim, and kill one another.

Posed another way, if *all* human beings had evolved at one central geographical location until relatively recently (perhaps within the last few thousand years), natural selection would not have had time to cause noticeable physical or cultural differences among us. So, for example, the result might have been that all human beings would now have black hair, blue eyes, brown skin, and more uniform height and features. Most importantly, being from one group, we would all speak the same language and share similar cultural and religious beliefs. Faces and personalities would distinguish one person from another, but there would be far fewer differences among us. In such a world, we would never have enjoyed the richness and fascination of our diverse cultures, and the course of human history would have been drastically altered. People might have found other reasons to discriminate and enslave in a monocultural world. But could we have avoided the thousands of wars among nations and the millions of deaths, injuries, and enslaved populations that resulted from cultural and religious differences?

We will never know the answer to that question. Yet, we could live in greater harmony by understanding human evolution and now knowing we are one species. We are all descended from the little lemur. Indeed, as we will see in Part Six, our first common ancestor is the one-celled animal that first appeared on Earth on September 25 in the Year of the Universe.

It is ironic that once we had reached the point in our evolution and development when our thought processes were sophisticated enough to discover the laws of nature, natural selection became far less of an influence upon our everyday lives and our long-term destinies. Since about 10,000 years ago, when we organized ourselves into cooperative communities, it has mattered little how fast any one of us could run. Humans now survive in a hostile environment solely because we can change the setting, not because we can adapt to it by natural selection. We inherited all our physical abilities and traits from our successful ancestral species. But now our species is defined almost solely by its ability to contemplate the future, form a self-image, and reshape the environment, for better or for worse—an incredible capacity that is due solely to the forces of natural selection that molded our brains into the most powerful force on this planet.

We no longer play by the rules of natural selection. We change our environment to suit us. When infectious organisms make us ill, we invent drugs to kill them. If we cannot see well, we put on glasses, and if we are too slow to catch a rabbit, we pick up a gun and shoot it. This behavior has allowed us to become increasingly weak, more prone to infection, and more dependent on our brainpower for survival. How long we will persist as a species is dependent on whether we can develop a plan to successfully maintain our environment. While the other billions of living creatures on this Earth provide the raw materials, niches for survival, and the pressures on natural selection that maintain the natural balance in our world, we are the one organism that understands these complex processes, yet we are forever radically altering them.

Although we now understand the process of evolution in great depth, we continue to see emotion swirling around it. In an effort to justify their irrational view of evolution, many people claim it is still a "theory." Indeed, in 1859 it was a theory. Today, however, it is an absolute and inescapable principle and fact. Calling evolution a theory—thereby ignoring today's incredibly solid and powerful edifice of information and knowledge—is nothing more than a simplistic word-game, an issue of semantics. Calling it a theory is perhaps the greatest misnomer in society and has nothing to do with science or scientific debate. Like thousands of other scientific facts proven beyond a reasonable doubt, whenever evolution is challenged by creationists, such "controversy" does not have two sides with roughly equivalent credibility. Very simply, as Huxley told Wilberforce in their debate in 1860, since humans are "highly endowed by nature," we don't do justice to our species and do not deserve this fantastic ability to learn and to reason if we "introduce ridicule into a grave scientific discussion."

Chapter Fifteen

Continental Drift

"The continents are adrift, ferried about on great fragments of the Earth's shell . . . The plates shudder and quake more than a million times a year. 'Solid Earth' is not as solid as we thought. It is energetic, dynamic, and fundamentally restless . . . It is continually creating its own surface, destroying it, repairing it, renewing it like a skin."

JONATHAN WEINER
Planet Earth (1986)

Well before Darwin wrote *The Origin of Species*, people were aware of the complex interrelationship of plant and animal life in ecosystems, and the effect of climate, topography, and other environmental factors on survival. With the advent of evolutionary theory, the view of that interrelationship was expanded to include natural selection. But a new and revolutionary discovery in geology beginning in the 1960's presented a critical aspect of the evolutionary process. At the same time that paleoanthropologists were collecting the fossils that enabled us to trace our ancestry back 65 million years, new geologic evidence was emerging explaining a startling and previously unknown cause for many of those external environmental changes and pressures that trigger natural selection.

With the San Andreas Fault in California as the dividing line, two giant land masses known as the Pacific Plate and the North American Plate (two of ten major plates of the Earth's crust and mantle that cover its surface) grind and slide past each other at about the same rate as our fingernails grow. In about fifteen million years, Los Angeles and San Francisco will sit directly next to each other. Six major fault systems underlie Los Angeles alone, and a hundred active faults pose a constant threat to this major population center. Though there is virtually nothing that can be done to diminish this threat, we now understand its fundamental causes and realize that it is part of a global phenomenon that has had a major impact on all life, not only the citizens of California, for millions of years. The discovery that the plates exist and our knowledge of plate tectonics—the dynamic processes of continental drift—has unified geology and evolution in a remarkable way.

Earthquakes and Volcanoes Give Rise to Mythology as They Cause Devastation and Death Throughout the World

Whales, Boars, and Snakes

Lisbon (Portugal), 1755: Violent shaking kills 60,000 people, as they are drowned from the tidal waves produced at the earthquake's center off the coast, crushed by the 12,000 buildings destroyed, and burned in the fire that lasts six days after the initial shock.

Laki (Iceland), 1783: Ashes from this volcano ruin vegetation, leading to famine and epidemics resulting in the death of 10,000 people.

Unzen-dake (Japan), 1792: Mud streams caused by this volcano destroy villages on its slopes, killing 10,452.

New Madrid (Missouri), 1811: The first of a series of earthquakes and aftershocks that lasts for nearly two months are felt in a radius of a thousand miles, and cause widespread flooding and damage.

Tambora (Indonesia), 1815: More than 12,000 die from this volcano's shower of rock and ashes, which also destroys crops and wildlife. The resulting famine causes the death of 80,000 people on the neighboring islands of Sumbawa and Lombok.

Krakatoa (Indonesia), 1883: After having been dormant for more than 200 years, ash-laden clouds spew forth from this volcano beginning on May 20, reaching a height of six miles. The volcano is active intermittently over the following months, until it climaxes on August 27 with perhaps the most violent eruption in modern history. The explosions are heard 2,200 miles away in Australia, ash is propelled fifty miles into the atmosphere, and the eruption triggers tidal waves that kill 36,000 people in nearby Java and Sumatra.

Mount Pelée (Martinique), 1902: A volcanic cloud of ash and lava kills 29,000 people as it sweeps over the town of Saint-Pierre.

Kelud (Java), 1902: When this volcano erupts, the mudflow and flood resulting from the release of waters from the crater lake causes the death of 5,100 people.

San Francisco, 1906: On April 18, at 5:12 a.m., the San Andreas Fault slips along a segment several hundred miles long, resulting in the death of 700 people and causing a fire that destroys the central business district of the city.

Tokyo, 1923: A violent earthquake causes an estimated 140,000 casualties, as nearly half of all the structures in the metropolitan area collapse.

Chilean Coast, 1960: Over a distance of 1,000 miles along the coast of Chile, 5,700 are killed and 3,000 are injured from this quake. The resulting tidal waves in Hawaii, Japan, and the Pacific coast of the United States cause further death and destruction.

T'ang-shan (China), 1976: On July 28, an earthquake near this coal-mining and industrial city east of Beijing causes the most devastating natural disaster in modern history—240,000 people are killed and 500,000 are injured.

According to the mythology of the Mayawyaw tribe in Luzon, Philippines, a goddess by the name of *Aninito ad Chalom* was the cause of all earthquakes. In the legends of the Bhuiya tribe of India, the goddess *Baski Mata* stood on her head at the bottom of the sea, supporting the Earth on her feet. Whenever she tired and shifted position, there was an earthquake. The primitive people of Chile believed a monstrous whale lived inside volcanoes. In India, the cause of volcanoes was thought to be a gigantic mole or a ferocious boar. Indonesians believed that when the snake *Hontobogo* moved, the Earth shook and fire erupted from the mountains. Animals were sacrificed in Sumatra to appease the spirits that dwelled in volcanoes. The Bagobos tribe in the Philippines sacrificed humans. Myths about Cyclops and Titans were inspired by volcanic activity.

Aristotle believed that volcanic explosions and earthquakes were caused by hot winds that move underground and occasionally burst forth in volcanic activity and earthquake tremors. He believed that these hot winds were caused by compression that was created when waves breaking on the seashore forced such air into caves that exist along the coast. Like many of Aristotle's other theories, this "pneumatic theory" endured for over 2,000 years, only to be ultimately disproven.

In March 1996, the previously dormant volcano "El Popo" (for "Popocatepetl," Aztec for "smoking mountain") forty miles southeast of Mexico City suddenly came to life and began belching ash and steam. Indians in the village of Xalitzintla, and many more among the 400,000 people in El Popo's shadow, trekked up the ash-coated slope of the 18,000-foot volcano with offerings of fruit, flowers, and chile sauce, in the hope it would not erupt and bury them in lava. Primitive myth still flourishes as current reality in many cultures—to make the world intelligible and gain influence over it, to ward off disasters, to reduce the fear and escape from the powerlessness, insecurity, and confusion in the face of overwhelming physical phenomena such as earthquakes and volcanoes. Fear first created the gods—fear that their village would be next, that it would be buried in ash or shaken to its foundations. In light of the history of death and destruction caused by earthquakes and volcanoes, it is no wonder that these once mysterious activities generated so many attempts to explain them and became an integral part of cultures around the world.

Beginning with James Hutton and Charles Lyell, geologists through the 19th and into the 20th century made inroads into understanding the causes of earthquakes and volcanoes. However, as the story unfolded in the mid-20th century, geologists began to realize that truth was stranger than the fictional mythologies that centered on these activities. Though Hutton and Lyell recognized a constantly changing Earth, they never imagined what modern geologists would eventually discover about the true cause of these phenomena.

Massive Plates of the Earth's Crust Gradually Redraw the Continents

Pangenesis and the Island of India

The theory of continental drift was originally proposed in 1912 by the German geophysicist and meteorologist Alfred Wegener (1880–1930). When he studied the similarity of the coastlines of eastern South America and western Africa, he speculated they had been joined together until about 250 million years ago as part of one giant continent called Pangaea (*pan* meaning "all," and *gaia* meaning "Earth").

Contrary to other theories about shifting continents, Wegener's

Figure 15-1

PANGAEA

Pangenesis theory stated that the continents had broken off of Pangaea and slowly drifted thousands of miles apart, like pieces of a cracked ice floe. In his 1915 book, *The Origins of Continents and Oceans,* Wegener noted that the Appalachian Mountains appear to be part of the same line of mountains that goes up through Nova Scotia, then down through Scotland and Scandinavia; that an east-west mountain range in Argentina matches one in South Africa; and that a plateau in Brazil fits into a similar structure in Africa's Ivory Coast. The examples he relied on were not limited to the physical similarities. Fern fossils at certain locations around the world also matched to such a remarkable degree that they alone could justify the Pangenesis theory. Despite his substantial evidence, by the time Wegener lost his life on an expedition in Greenland in 1930, most geologists discounted his theory of continental drift. Pangenesis sank into obscurity for decades. As late as the mid-1950's, college geology textbooks reflected our lack of knowledge about mountains, volcanoes, and earthquakes. For example:

> The cause of the . . . mountain ranges is not yet established . . . Until a much more complete knowledge of . . . the interior of our planet is obtained . . . the origin of mountain ranges and volcanoes will remain unsolved . . . The positive geological results attained by seismological methods . . . foreshadows a new and exceedingly interesting era of . . . research.

Indeed, the science of geology was on the verge of an enormous advance. As a result of extensive exploration of the seafloors to track submarines during World War II, combined with other geologic evidence that began to mount in the 1950's, a more detailed and accurate version of the theory of continental drift evolved into the modern field of plate tectonics. The evidence is now overwhelming. The new view of planet Earth is primarily the result of these discoveries:

- *Similarities in fossilized organisms* in strata of distant continents along areas where they would physically fit (such as South America and Africa).

- *Similarities of rocks and structures* of distant continents in those same areas.
- *Consumption of the Earth's crust* in areas around the world, such as the northern edge of the Australian Plate being forced under the Eurasian Plate (forming the Himalayas), and the western edge of the Pacific Plate being forced under the Asian Plate, resulting in the proliferation of volcanoes and earthquakes in Japan and the surrounding area.
- *Configuration and composition of ocean floors*, such as the division between plates at the 46,000-mile-long mid-oceanic ridge, caused by the generation of new crust boiling out of the Earth's mantle. This phenomenon is known as seafloor spreading, and is determined by age-dating of minerals and sediments that make up the ocean floor.
- *Direction of rock magnetism in ocean floor* as shown by "striping" that exists in mirror-image fashion, tracing its "growth-lines" to expansion of ocean plates.

The direction of magnetism of minerals in the ocean floor has provided fascinating and conclusive evidence of seafloor spreading. For many years we've been able to measure the intensity and direction of the Earth's magnetic field with instruments called magnetometers, towed behind ships. We also know that the polarity of this field changed naturally about 700,000 years ago, and that at the time rock is formed at the Earth's crust, including at the ocean floor, the iron molecules in the rock line up in the same direction as the Earth's geomagnetic field in which the rock exists. The measurements by magnetometers revealed 180° deviations in certain rocks that comprise the ocean floor, particularly at the crests of the mid-oceanic ridge. That is, intermittent portions or strips of sediment had the exact opposite polarity or magnetism from the strips next to them. In 1963, two British geophysicists, Frederick J. Vine and Drummond H. Matthews, proposed that the deviations in magnetism found in these sediments resulted from volcanic rock steadily emerging from beneath the Earth's crust and spreading in both directions as the Earth's magnetic field reversed itself. As they matched the magnetic rock of varying width and magnitude on either side of the great oceanic ridge, these parallel strips formed mirror images of each other, much like matching up rock strata at distant locations. This is illustrated in Figure 15-2.

Figure 15-2

SEAFLOOR SPREADING AT MID-OCEAN RIDGE

Plate A ← **▲** → Plate B
Molten rock
pushes up

The pace and progression of seafloor spreading is one to five inches a year, depending on the particular location. This information on magnetism of the rock revolutionized concepts of the origin and age of the ocean floors, and confirmed the phenomenon of seafloor spreading, which in turn was instrumental in resurrecting Wegener's original theory about Pangaea and continental drift. Alfred Wegener didn't have available to him the sophisticated equipment, detailed tests, satellite images, and computer modeling that can now explain the Earth's internal dynamics and trace its history. But after decades of refining global theory, we now see how continents can actually "drift." Each plate is made of the Earth's crust and the top portion of the next layer of material, called the mantle. Each plate ranges from a few miles to as much as sixty miles in thickness, and "floats" on the rest of the remaining portion of underlying mantle, which is molten. As shown in Figure 15-3, the plates' edges have no particular correlation to the seven continents or to shorelines since the location of oceans and seas, of course, is due to land elevation, not the plates' shapes or boundaries. If we entered another ice age, gla-

Figure 15-3

CROSS-SECTION OF PLATE/CONTINENT

ciers and polar caps would enlarge, causing sea level worldwide to drop, and the portions of land above sea level would increase. If, on the other hand, the Earth's temperature were to increase over the next few thousand years, the polar caps would melt and much more of the presently exposed land area of the Earth would be covered by water. Movement of the plates themselves has also changed shorelines dramatically back and forth across the continents over the past tens of millions of years, apart from global changes in the ratio of ocean to ice. Thus, the label "continent" is merely the identification of that 25-percent portion of the Earth's plates that happens to be above sea level at this particular point in the Earth's history. "Continental" drift is a misnomer since it is actually the *plates* that are drifting, with the highest portions being exposed to air as they ride about on those largely submerged plates in a dynamic process by which the Earth's surface is continuously redesigned. At one edge of each plate, fresh molten rock flows up from the Earth's mantle and causes the plates to push apart. This is the seafloor spreading shown in Figure 15-2. At the other edge of the plate, old rock is pushed down into deeper zones where it is melted down again, or collides directly against the opposing plate's edge. Fifty million years ago, India was an island, slowly moving toward Asia. Though India is now part of the Asian continent, it's actually the northern tip of the Australian Plate, which was pushed higher when it collided with the Eurasian Plate (see Figure 15-4).

The Himalaya Mountain Range welled up along the line where these land masses meet. Over the last 40 million years, Mount Everest, made of material that once lay on the calm sea bottom, was thrust up to become the highest point on Earth. Similarly, in early

Figure 15-4

MIGRATION OF INDIA INTO THE ASIAN CONTINENT, INCLUDING EARTHQUAKE FAULTS

India Today

India
40 million years ago

1996 researchers found rock high in the Swiss Alps that had once been 400 miles deep in the Earth's mantle and had been thrust upward as a result of continental collisions like the one now taking place between the Australian and Eurasian Plates. North America, like many other continents, is a conglomeration of pieces and blocks of plates that broke from larger plates and drifted together over the last 200 million years.

Earthquakes and volcanoes proliferate along the areas where the plates rub and chafe past each other in slow motion around the world. The outline of most plates is evident from the number of earthquakes and volcanoes at their edges, as experienced firsthand by the residents of Japan (which is shaken a thousand times a year),

Figure 15-5

WORLD MAP OF MAJOR PLATES, IDENTIFYING DISTRIBUTION OF EARTHQUAKE EPICENTERS (BLACK AREAS) RELATED TO PLATE BOUNDARIES

(1) Eurasian Plate

(10) African Plate

(9) South American Plate

(8) North American Plate

(7) Caribbean Plate

(6) Nazca Plate

(4) Cocos Plate

(5) Antarctic Plate

(3) Pacific Plate

(1) Eurasian Plate

(2) Australian Plate

California, and other places around the world that happen to strad-
dle these massive and dangerous cracks in the Earth's surface. One
million quakes or tremors shake the Earth's surface every year.
About 50,000 earthquakes are measured with seismic instruments
every year, and about a hundred of these are large enough to cause
significant damage. Tokyo has been laid to ruin by earthquakes an
average of once a century for the last 2,000 years. It is not surprising that Japan spends about $100 million each year on earthquake
prediction research.

Ironically, Japan owes its very existence to volcanoes caused by
the same series of phenomena that cause the earthquakes. The East
Pacific Rise (the counterpart of the Mid-Atlantic Ridge) is the result of seafloor spreading that pushes the Pacific Plate toward Asia.
However, as the Pacific Plate advances, it is sliding under the
Eurasian Plate and into the Earth's interior, resulting in tremendous friction that causes some of the most violent earthquakes and
volcanoes on Earth. The islands that make up Japan have risen out
of the sea over the last 30 million years as the direct result of the
volcanic activity at this juncture of plates. As recently as April 1973,
volcanic eruptions south of Tokyo caused lava to rise above sea level
to form the new island of Nishimo. Similarly, the underwater volcano Loihi, 20 miles southeast of the island of Hawaii, is in its 14th
year of continuous eruption, and its peak—which is now 3,000 feet
below the ocean surface—will form the next Hawaiian island in
about 100,000 years.

The forces driving this ponderous movement originate in the
Earth's core and mantle. The iron core is 1,500 miles in diameter,
has a pressure 3.6 million times the pressure at the Earth's surface,
and has a temperature of about 12,000°F, compared to the Sun's
surface at a mere 10,400°F. As this tremendous heat is released, it
causes convection and massive disruptions in the Earth's mantle.
The mantle, the dense layer immediately below the Earth's lighter
crust, is about 1,800 miles thick. The rigid, topmost forty miles of
mantle, together with the crust, form the tectonic plates. As one
goes deeper, the layers of the mantle are being heated by the core,
and churn like thick soup in slow motion. The earthquakes and vol-

canoes are the superficial results of the grand motion and rise of heat from the inside of our planet.

Life Adapts as Plates Drift

Two Months Without Food . . . Life in Total Darkness

As these massive plates slowly dance with each other, all of life is affected. The plant and animal life that propagated across the Earth on the continents and in the seas has been riding the plates for hundreds of millions of years. The plates carried life with them from one location on this sphere to another, like a raft slowly moving downstream. Some plates began in a cold and dry region and end up in a tropical environment, while others have applied their tremendous force against each other, resulting in the crust buckling into the world's massive mountain ranges. There are unmistakable remnants of tropical jungles in Alaska, and the marks of great glaciers beneath the Sahara. Plants and animals adapted as these rafts made their way along the route. The land beneath them rose and fell over millions of years, as natural selection dictated the physical and mental characteristics favorable to survival of each species. Thus, plate tectonics has profoundly affected the history of life on Earth.

Great spans of time are prerequisites for such adaptation. This is seen in what could be called the "Iron Turtle Competition"—the remarkable history of the green sea turtles that migrate 1,250 miles from Brazil to tiny Ascension Island in the middle of the Atlantic Ocean. Why don't they stay in Brazil and nest on the hundreds of miles of beaches there? What explains the choice or "programming" of these 400-pound reptiles to undergo this impossible trek—without food and in the face of a myriad of dangers? How do they even know the island is out there? Only one explanation makes good sense. It is based primarily on the fact that tens of millions of years ago there were islands only a stone's throw from South America. The ancestors of these turtles made the short swim to nest and lay their eggs at a location where the eggs and young would be secure from the many predators on the mainland. As the sea floor spread and Pangaea separated into Africa and South America, the ocean expanse gradually increased between Brazil and the series of volcanic islands that have come and gone in the vicinity of Ascension. From one generation to the next, the turtles unconsciously adapted to the increase of a few inches a year, caused by the lava welling up between the plates. But as they adhered to a migration pattern that seemed identical with the tradition of their ancestors, they were actually committing their progeny to this incredible task millions of years later. Thus, the green sea turtle evolved into a larger animal and a stronger swimmer in order to survive and produce offspring. It developed heavy fat deposits and a metabolism that would enable it to travel 1,250 miles over a two-month period *without food!*

As mentioned earlier, lava erupts along portions of the mid-oceanic ridge. A variety of plant and animal life has evolved in these deep undersea "oases," just discovered in 1973 by scientists in small submarines researching the rift valley at the center of the Mid-Atlantic Ridge. Giant mussels, clams, crabs, fish, giant tube worms, and unique plant life flourish here, yet are totally dependent on the heat and minerals being spewed forth by tectonic activity. A unique "pocket" of natural selection, these vents that originated in the mantle are the only places on Earth where life is not directly or indirectly a product of sunlight. Instead, the plant and animal life at these depths, where there is no sunlight, consume bacteria that derive energy from hydrogen sulfide that results when the water mixes with the molten basalt beneath the ocean floor. Thus, in a relationship even more direct than the one experienced by the "iron" turtles, the creatures that live along these undersea rifts owe their very existence to plate tectonics.

Plate tectonics has also been a direct cause of reproductive isolation (discussed in Chapter 13) ever since Pangaea began to break apart. We've already seen numerous examples of reproductive isolation—from the differences in the finches' beaks and tortoise shells in the Galapagos Islands to the unique physical traits among the races of humans that used to be geographically separated. We continue to discover previously unknown effects of plate tectonics. In the fall of 1995, a team of French and British scientists on an expedition deep into the unexplored Riwoche mountain region of northeast Tibet entered a 17-mile-long icy valley, isolated by 16,000-foot-high passes on both ends. They discovered an entire herd of an archaic-looking horse that had never been seen or known before by biologists. The direct result of reproductive isolation caused by the formation of the Himalaya Mountains, this four-foot-tall animal had branched off over five million years ago. It has retained several unique and primitive physical characteristics that distinguish it from all other breeds of horse, such as its small stature and triangular head shape that resemble the images of the vanished horses seen in Cro-Magnon cave paintings.

One sweeping global theory, plate tectonics, has fused broad perspectives and diverse information discovered by geologists, biologists, and physicists. These scientists have now accounted for the mid-oceanic ridges, the world's great mountain ranges, almost all of this planet's earthquakes and volcanoes, the birth and disappearance of oceans, the movement of continents, and an essential chapter in our knowledge of evolution.

Part Six

The Cell and Genetics

In Focus

Eleven billion years after the Big Bang, a handful of elements
combined on this Earth to begin an incredible process. Life emerged
from the primordial soup. Life from lifeless electrical charges. Life that
can now be explained in terms of physical constituents and processes,
even though such explanations might seem too cold and dispassionate
to account for the richness and complexity of living things and human
consciousness.

The origin of life on Earth, reproduction of living things, and
Darwin's theory of evolution were stripped down to their physical and
chemical basis in the 20th century—down to the point where we were
able to observe the previously undetectable factors of inheritance, the
mainspring and mechanism of evolution. For the first time, we could
understand the interrelationship of those three elements of life: origin,
reproduction, and evolution.

First, we must consider the cell and genetics examined in this Part
Six, then to the blueprint and backbone of life in Part Seven. From the

cosmic egg in Part Four we now explore the human egg. From the cosmos to the chromosome. From the stark and bare expanse of Hubble's galaxies, to the infinitesimal, the microscopic, the molecular, the unseeable, the unimaginable within that is sustaining us. From the physicists' atomic energy to the biophysicist's energy of life. From evolution's macro view of life to the biochemist's micro view.

The story of cells, genes, and the molecule of life in Parts Six and Seven explains how thousands of people united the knowledge accumulated in diverse fields into one cohesive resolution, beginning with the discovery of cells in 1673 and the meticulous labors of an obscure monk at his monastery in the former Czechoslovakia in the 1860's.

Chapter Sixteen

Primordial Soup

"Biology . . . is primarily a descriptive science, more like geography, dealing with the structure and working of a number of peculiarly organized entities, at a particular moment of time on a particular planet."

JOHN DESMOND BERNAL
Science in History (1965)

Darwin had accumulated volumes of data and examples to support his theory of evolution. Evidence and observations accumulated by scientists in the succeeding decades gave more support to the validity of evolutionary theory. Yet despite all this, the theory lacked evidence from *within* the living organism. Thus, Darwin's theory of evolution begged this question: *What is the mechanism by which this natural selection works?* Exactly what is happening inside these living organisms to allow evolution to take place?

Well before *The Origin of Species*, biologists and others also attempted to understand reproduction of living things, and what directs, determines, and controls their growth, as we saw in Chapter 12 with the theories of spontaneous generation and preformation. The discoveries made in biology in the 19th century, together with evolutionary theory, set the stage for unraveling the mystery held deep within us and answering the most fundamental questions of life. *How did life first arise on Earth? How do organisms reproduce? What is the internal physical basis for evolution?*

The Cell, Its Nucleus, and Its Process of Dividing Are Discovered

Animalcules and Crucial Signals

What the atom is to physics, the cell is to biology. The atom is the building block of all matter. The cell is the building block of all life. Despite the incredible diversity of living things in this world, there is an underlying unity. All cells work essentially the same way, and the genetic code that governs all life forms is identical in all species—conclusive evidence that all forms of life share a common ancestor, as Darwin had predicted in *The Descent of Man* in 1871, well before anyone understood the role of the cell in heredity or reproduction. Now we have discovered that each cell is a tiny chemical factory that is capable of processing its own nutrients, generating its own energy through the use of those nutrients, communicating with neighboring cells, and *dividing* into two identical units. *The cell's ability to replicate itself is the key to all life and growth.* A single cell can be a complete organism in itself, such as a bacterium or amoeba. Other cells are specialized as the building blocks of multicellular organisms.

Though the average-sized cell contains trillions of atoms, it's still much too small to be seen with the naked eye—it would take about 10,000 human cells to cover the head of a pin. So, no one imagined the existence of cells before they could be observed under the microscope. In his 1665 publication, *Micrographia* ("Small Drawings"), the English scientist Robert Hooke (1635–1703) coined the word "cell" before any actual living cell had been seen. He used the term to describe dead plant tissue and the microscopic honeycomb cavities in cork, because the cellulose walls of the cork reminded him of the blocks of small rooms or "cells" occupied by monks in a monastery. Single-lens magnifiers were first developed in the mid-15th century but weren't powerful enough to see *living* cells of animals until Antoni van Leeuwenhoek (1632–1723) produced more powerful lenses in the 1670's. In 1673, van Leeuwenhoek opened a whole new world when he discovered blood "cells" (adopting Hooke's word), spermatozoa, and single-celled organisms such as bacteria and swimming protozoans. Much like Galileo's first observations of stars in 1609, van Leeuwenhoek's initial glimpse of these cells and one-celled organisms was merely a small step in a journey that would unfold over the next three centuries. Neither man had any idea how immense and significant that journey would be, only that they could see objects that were not visible with the naked eye and that were unimaginable prior to those observations.

Much like 17th-century astronomers' efforts to label and mark the locations of newly discovered stars, and similar to Linnaeus's system of classifying animals in the 18th century, van Leeuwenhoek's discovery of what he termed lively "animalcules" first led to categorization of one-celled animals. In subsequent years, doctors and biologists categorized various types of tissues (mucous, fibrous, etc.) by viewing their cells under the microscope. For example, cells that make mucous were found to make up the linings of the airways and gastrointestinal tract, and those making fibrous connective tissue constitute the layers under our skin and surrounding our muscles. Three different kinds of muscle cells were discovered (cardiac, involuntary, and voluntary), as well as numerous specialized cell types that are components of the nervous and reproductive organs. However, even as this labeling effort progressed rapidly through the 18th century, the functions and contents of cells remained a mystery.

Because the early microscopes were not strong enough to view all the details within cells, and because of the persistence of some of the archaic theories of spontaneous generation and preformation, cell theory didn't begin to develop until 1831. In that year, Robert Brown (1773–1858), a Scottish botanist, observed the control point of the cell, calling it the "nucleus," and identified that structure as being a common element of all plant cells—a discovery that rivals the importance of the later discovery of the atomic nucleus. Nuclei were soon discovered in animal cells, and the flowing of "protoplasm" (internal material in the cell) was observed in living cells in 1835. Two German biologists, Matthias J. Schleiden (1804–1881) and Theodor Schwann (1810–1882), refined and developed this early information into the first basic principles of cell theory. In 1838 and 1839 they wrote that all cells are composed of membrane, nucleus, and cell body and are the elementary particles of all plants and animals.

Many theories about the "birth" of new cells developed during the 1840's but were discarded in favor of cell division, the process by which two identical "daughter" cells result from the original. "All cells come from cells," asserted the famous German physician and biologist Rudolf Virchow (1821–1902). In his 1858 book *Cellular Pathology*, Virchow explained that life is not a result of supernatural phenomena and that "cells are the link in the great chain of . . . formations that form tissues, organs, systems and the individual." This established the modern concept of cell pathology, that is, the study of disease processes beginning at the cellular level. After Virchow and others proved the importance of cell division, cells became the subject of intense research, soon leading to the general acceptance of the fact that living things can only arise from the seeds of their parents. Life is never created anew on Earth, but is descended from ancestors in an unbroken line. Coincidentally, Virchow's book was published in 1858, the same year that the joint paper by Darwin and Alfred Russel Wallace was read to the Linnean Society, and a year before publication of *The Origin of Species*. However, it would take generations before evolution and cell theory would converge in a meaningful sense.

Biologists Discover the Cell's Organs

The Cell's Brain, Stomach and Skeleton

Over the succeeding decades, scientists began to identify other parts of the cell's internal machinery. Cells actually contain a set of distinct organs called organelles, not one "protoplasm" substance as once thought. Organelles are *ultramicroscopic* (that is, they can be seen in detail only with an electron microscope) and are internal compartments of various shapes, sizes, and complexities, containing sets of specific chemicals. Organelles carry out chemical reactions on materials such as nutrients, growth factors, and toxic agents entering or leaving the cells.

- *Nucleus:* Taking up about one-tenth of the cell's volume, the nucleus is the cell's "brain." It contains the genes and the nu-

Figure 16-1

ORGANELLES

Cross-Section of
Typical Animal Cell
(In Interphase, see Figure 16-3)

Mitochondrion

cleic acids (DNA and RNA) that determine the proteins that control all the chemical processes in the entire organism. RNA molecules, which carry the gene's instructions, are moved out of the nucleus (through its membrane) to the cell's protein-making organelles, known as ribosomes.

- *Ribosome:* These number in the thousands and sometimes millions in *each* cell, depending on how much protein the cell must make. Ribosomes are the attachment sites for the specific "messenger" RNA sequences that determine the proteins made by specific cell types.

- *Endoplasmic Reticulum:* The "ER" is a floppy bag of folded membrane that takes up about half of the cell's volume. Its function is to provide a separate chamber in which the proteins made on the ribosomes can be put in their final and correct chemical form. This is where sugars, phosphates, and other molecules are added to the proteins at specific chemical binding sites.

- *Golgi Apparatus:* A typical cell has thirty or forty of these smooth, membranous structures. It is the only organelle named after a person—Italian biologist Camillo Golgi (1844–1926), who discovered it. It stores and processes fats and performs even further modifications of proteins as well as preparing them for transport to other parts of the cell or out of the cell entirely. For example, proteins that are digestive enzymes made in the liver need to be shipped to the stomach. In the Golgi, specific sugars attached to these digestive protein molecules provide the information that directs the protein to its final destination.

- *Lysosome:* Using powerful enzymes, these organelles form the cell's multiple "stomachs" or digestive sacs. These enzymes break down the food we eat into its smallest components, which then are used in the various cells. For example, lysosomes break down the proteins we ingest into amino acids, which are reused by the cell to make new proteins.

- *Cytoskeleton:* These networks of ultrathin protein filaments maintain the cell's structure and shape, allow it to move, and support the cell's internal transportation network that links organelles and ferries products between them.

- *Mitochondrion:* These sausage-shaped structures of folded membranes extract the energy from the chemical bonds of carbohydrates, fats, and proteins and convert that energy into a form that can be used by the other organelles to replenish the energy they expend in conducting their work. Amazingly, mitochondria have their own DNA and their own ribosomes, and they divide *within* cells. These characteristics, together with the fact they are shaped like bacteria, indicate that mitochondria were once bacteria that originally entered cells as parasites and later evolved into this unique organelle.

- *Membrane:* The cell membrane (also called "outer plasma membrane") not only holds all the cell's organelles inside, it vigorously controls the entry and exit of molecules that are critical to cell function. The molecules that comprise the membrane are lined up unerringly with a "head" that is "hydrophilic," meaning it attracts other heads and is compatible

Figure 16-2

CELL MEMBRANE STRUCTURE

Hydrophilic heads
(compatible with
fluid environment)

Lipid (fatty)
bilayer

Hydrophobic
(hydrocarbon) tails
(avoid water)

Protein molecule
senses external signa
and communicates tc
inside cell

with the watery environment in which cells reside, and a "tail" that is "hydrophobic," meaning it repels "heads" and water. This configuration allows for proteins to be embedded within or to span the entire membrane. Some of these proteins are actual channels that control what the cell takes in and sends out, while others form receptors that detect specific external signals, allowing the cells to respond to stimuli and conditions.

Each cell contains only one nucleus, while the other organelles are present in multiple units. In addition to the organelles listed above, plant cells contain chloroplasts, which are much like mitochondria, but create energy through photosynthesis.

Biologists Determine the Phases of Cell Division

Dying of Old Age

Cell research reached an important phase with Walther Flemming's work in the 1870's. Flemming (1843–1905), a German physician and anatomist, improved the dyes used to see cell structures. In 1879, he was able to identify a threadlike material in the nucleus of cells. Observing that material (later called chromosomes) during cell division, Flemming showed that the threads shortened (i.e., contracted and condensed) and split longitudinally in halves, with each half moving into opposite sides of two new identical cells. He called this cell-division process "mitosis" and described it in his 1882 book, *Cell Substance, Nucleus, and Cell Division.* He thereby founded cytogenetics, the study of the physical processes involved in inheritance at the cellular level.

After Flemming's work we gradually began to understand the physical processes within the cell. Yet, it took until the middle of this century to figure out what is contained in the cell and its nucleus that enables the cell to control its own functions and the functions of an entire organism. The "master plan" by which these cells undertake the complex steps to create a new single cell that proceeds to divide into two, four, eight, and ultimately sixty trillion cells that comprise a complex living organism such as a human will be examined in Part Seven. Like everything else in the world, the physical process of each cell works according to the established laws of chemistry and physics. This includes the two basic steps of the reproductive process—cell growth and cell division (mitosis).

A major portion of the cell's machinery is dedicated to the process of mitosis. This is because multicelled plants and animals must be able to replace cells that die naturally or are injured, and single-celled organisms must reproduce themselves. Regardless of whether we are considering a liver cell or a lung cell of the human body, a plant cell or a free-swimming protozoan, the process of cell division is essentially the same because the ultimate goal of the process is to produce new cells exactly like the ones from which they originated. Figure 16-3 depicts the process of mitosis in a normal cell. For purposes of illustration, this figure shows only two pairs of chromosomes in the cell's nucleus. However, the nucleus of every human cell actually contains 46 chromosomes (23 pairs), except the egg and sperm cells, which contain 23 chromosomes. Also, red blood cells have no nucleus, thus, no chromosomes.

A cell's life consists of the following five stages:

Figure 16-3

STAGES OF MITOSIS

1. Interphase
(Cell preparing
for mitosis)

2. Prophase
(Chromosomes form
in nucleus, nuclear
envelope breaks down)

3. Metaphase
(Chromosomes
line up at "equator"
in nucleus)

4. Anaphase
(Chromosomes
separate)

5. Telophase
(Two new cells
separate and nuclear
envelope reforms)

Resulting in
two identical
"daughter" cells

1. *Interphase:* The cell spends most of its life in this phase. It is
   not replicating during interphase, but is carrying out its essential functions, including making specific proteins, secret-
   ing "lipids" (fatty, water-insoluble compounds), and
   manufacturing the connective tissue matrix that holds us together. The pace of these activities depends on what the body
   needs at a given point in time. For example, every time we eat
   a meal, the cells that line our stomach and intestines must
   manufacture large amounts of proteinaceous enzymes and

acids to digest the food. The cells making these materials are in the "interphase" of the cell cycle, but they are also fully dedicated to this secretory function and will live only weeks to months, soon to be released into the gastrointestinal tract at the end of their useful lives. Cells receive the signals to divide during interphase.

2. *Prophase:* A cell enters prophase after receiving an external stimulus to divide while in the interphase. During prophase, the long coils of DNA that are wound diffusely through the cell's nucleus begin to condense into organized structures called chromosomes. Thus, it is preparing to divide. Its neighboring cell may have died of old age, or other cells in the population may be under attack by toxic agents or infectious organisms. Whatever the reason, the goal is to produce two new identical cells. To do so, the chromosomes of that cell (thus, the DNA containing the vital genetic information) must be duplicated. The duplication of DNA—the result of a complex set of biochemical reactions—will be examined in Part Seven.

3. *Metaphase:* The defining characteristic of metaphase is the ordered lining up of all the chromosomes across the central "equator" of the cell, so that they are in a position to move equally to one new cell or the other. This critical event is controlled by precisely laid "cables" that will guide the chromosomes to their respective cells. The cables are part of the cell's cytoskeleton, described above—the dense array of microscopic filaments and tubes that dictate the cell's shape and allow its movements. The nuclear membrane that was so prominent during interphase and much of prophase dissolves and no longer defines the nucleus. The entire function of the cell is now dedicated to a successful division into two new cells. Once metaphase begins, the cell division process (including the final two stages, anaphase and telophase) typically lasts only a few minutes.

4. *Anaphase:* This marks the beginning of chromosomal separation, with the newly copied chromosome and its original having moved apart. The chromosomes are now on their way to

the developing cells. Chromosomal movement was thought for many decades to be a function of the filaments "pulling," but new information has shown that the chromosomes actually migrate along the cables by a type of "ratchet" mechanism.

5. *Telophase:* The chromosomes have now reached their respective cells, and the two new cells are forming a new membrane between them so they can perform their functions and reside side by side, such as in human tissues, or, for example, so they can swim away as two new protozoans. The pairs of chromosomes soon become enclosed in a nuclear membrane and the cell moves rapidly toward interphase, beginning the life cycle of the cell once more.

What causes each phase of mitosis? As a result of discoveries made primarily during the 1970's and 1980's, molecular biologists now understand the crucial events that trigger cell growth and cell division to an amazing level of detail. Complex chemical processes control each phase, ultimately directed by the DNA molecule, which contains the master plan of all life. But before looking at how sixty trillion cells, such as those comprising one person, can possibly function as a single organism, we will look at one cell—the origin of the very first living cell on Earth.

Life Springs from the Primordial Soup

Vents and Tube Worms

There are no seas on this planet because it is so hot that water would vaporize. Most of its surface is molten rock; the crust that does exist is tenuous, brittle, and fleeting. Meteorites plunge through a thin atmosphere of hydrogen and helium into a surface mix of methane, ammonia, hydrogen, carbon monoxide, carbon dioxide, and nitrogen. Along with the meteorites, intense ultraviolet radiation from the sun penetrates to such a high degree that a lethal dose for any living organism is delivered in less than an hour. The gravitational pull of this planet's moon—only 11,000 miles away—raises massive tides of molten rock.

This is not Jupiter or a planet in some other solar system. This is Earth 4.6 billion years ago, soon after it coalesced from dust and gases left over from the Big Bang into a shimmering sphere of molten material. This boiling, radiating inferno seems an unlikely location for the beginning of life. But over the next several hundred million years, the bombardment of meteorites slowed and the tremendous heat radiated out into space, allowing the surface to cool and stabilize. An atmosphere of heavier elements began to form, resulting in the formation of a layer of ozone in the upper atmosphere, which blocked the sun's harmful ultraviolet rays.

The first billion years of the Earth's existence are called the Hadean Eon, named after "Hades," Greek for "hell." As stated by Jonathan Weiner in *Planet Earth*, "Life arose almost as soon as the planet ascended from hell." The first traces of life in the geological record are about four billion years old: simple one-celled organisms. What took place in the primordial soup to create what we now call "life"? As we will see in Part Seven, all life is made of complex molecules called amino acids, the building blocks out of which all plants and animals formulate their proteins. We will also see that molecules called ribonucleic acids (RNA) are essential for the existence of life because they "organize" those amino acids into proteins. However, amino acids and RNA molecules are themselves made by living things. That is, they are the product of complex workings within our cells, thus creating a "chick-egg" riddle. How could life have developed from amino acids and RNA if those substances are the *products* of living cells? The answer lies in the probable existence of both of those essential components in the primordial soup of four billion years ago.

The basic chemical elements that comprise amino acids and RNA were present at that time. In 1953 Stanley Miller, a graduate student at the University of Chicago in the lab of the famous chemist Harold Urey (who played a major role in the Manhattan Project), designed a series of large interconnected glass flasks intended to represent conditions on early Earth. In Miller's simulation, steam rose from boiling water and mingled with methane, ammonia, hydrogen, and water vapor, the major gases present in this planet's early atmosphere. Electrical sparks of 60,000 volts of electricity were then discharged through the mixture of gases to simulate lightning—all for the purpose of determining whether the critical amino acids could be created out of "whole cloth." After a week, Miller analyzed the chemicals in the water and found that it was rich with a variety of amino acids. This experiment has been reproduced countless times since 1953, always resulting in the formation of amino acids. Given the immense period of time and the billions of random combinations of chemical elements that must have taken place all over primitive Earth, these laboratory experiments provide a plausible explanation for the formation of amino acids, one of the molecules essential for life.

Likewise, scientists have demonstrated that the chemical constituents of RNA were abundant at the same time as amino acids were being formed. These basic elements (including phosphorus, nitrogen, hydrogen, carbon, and others) connected with one another to form long molecular chains by following the same physical principles of binding and valency that are followed by simple molecules. Certainly, some ribonucleic acids joined into chemical sequences that were meaningless. The mere existence of amino acids and RNA molecules does not explain how these building blocks eventually were configured and organized into living organisms. In order to "live," RNA needed two capabilities: (1) the ability to organize the amino acids into proteins; and (2) the ability to duplicate itself so it could continue to exist and cause that essential process to take place. As for the first capability, some RNA molecules lined up randomly in a configuration that caused specific amino acids to form primitive proteins. Those proteins could have promoted chemical reactions that would favor one RNA configuration over another. But RNA also has the incredible ability to promote such chemical reactions by itself, as demonstrated in the laboratory in 1982 by the biochemists and Nobel laureates Thomas R. Cech (1947–) and Sidney Altman (1939–). This "self-catalyzing" capability was an essential step in the formation of life. It resulted in evolution from a strictly "RNA world" to one in which RNA directed the manufacture of proteins. As we will examine in Part Seven, the making of proteins by DNA and RNA is the basis for all life.

Regarding the ability of RNA to self-replicate, even the simplest and most primitive form of RNA had the potential for its own duplication through the formation of an identical complementary RNA molecule due to the principles of chemical binding mentioned above. *Once formed, primitive RNA began the first version of natural selection.* If various forms of RNA had existed simultaneously, the ones that attracted the most advantageous pattern of amino acids and could multiply most rapidly would have had the best chance of proliferating and further monopolizing the essential chemicals needed for forming such molecules. The RNA molecules that accelerated favorable biological reactions survived. In this system, evolution of RNA took place. Paraphrasing Darwin, natural selection worked solely by and for the good of each *RNA molecule*. Over millions of years, RNA molecules evolved into distinct and more complex types, each with a specialization. Thus, RNA—the maker of proteins from amino acids—also had the consistent, overwhelming, and defining function of all living organisms, namely, self-replication.

Some biologists believe that the undersea vents at plate junctures (discussed in Chapter 15) were likely locations for the creation of life. At these depths, life was protected from the harmful ultraviolet rays. Also, essential elements and compounds such as carbon, hydrogen, methane, water, and ammonia were present at these sites, as well as a wide range of temperatures. With such a large number of locations throughout the world, like thousands of experimental labs, there was ample opportunity for the right chemical mix.

The formation of *membranes* around distinct groups of RNA and proteins assembled from amino acids was probably the next step toward cell development. As shown earlier, the lipid molecules that comprise the cell's membrane naturally line up in rows that form an interior and exterior sheet (see Figure 16-2). This compartmentalization most likely created the first cells and allowed them to exist in a variety of watery environments. Like the primitive RNA molecules, these first cell "entities" competed for common ingredients, leading to variations in cells and the survival of those most efficient in passing on to their progeny the catalysts and other characteristics that promoted duplication.

Bacteria Become the First Common Ancestor of All Life

Inner-City Homeless and the Genetics of Resistance

For the next *three billion years*, the only living things on Earth were one-celled organisms. Like the first RNA and the multicelled organisms that later evolved, these one-celled animals went through their own distinct evolutionary process. Cell biologists and biochemists have analyzed cell fossils and have been able to reconstruct the most likely evolutionary paths taken over those three billion years. Independent cells first lived by photosynthesis, making energy from sunlight. In the next major evolutionary hurdle, cells adapted from an atmosphere with virtually no oxygen to one that gradually became rich in oxygen over the 500 million years after cells first appeared. Cells that survived this increase in oxygen and began to use it to their advantage—a primitive form of respiration—multiplied to cover the early Earth.

These early cells evolved into bacterial species from which all other living things, including plant and animal life, evolved. Bacteria are one-celled creatures, and have all the organelles described earlier in this chapter, except a well-defined nucleus. It would take about 250,000 average-sized bacteria to cover the period at the end of this sentence. Figure 16-4 shows a typical bacterial cell. Some bacteria also have rigid hair-like projections (pili) covering their exterior, and others have a tail-like structure (flagellum) for mobility.

In favorable conditions, some species of bacteria are able to reproduce every fifteen minutes. Early ancestors of present-day bacteria entered into complex relationships with algae (primitive plants) and formed mound structures several yards high and across. These simple ecosystems, the fossils of which are called stromatolites, covered the shallow coastal waters of the Earth. The same type of formation is common today and is found in the shallow waters of the Gulf of California and along the Great Barrier Reef off the northeast coast of Australia.

Figure 16-4

Drawing of a Typical Bacterial Cell

Genetic material
(but no discrete nucleus)

Cell wall and
membrane

About 5,000 species of bacteria exist today. A relatively small number of bacterial species cause disease. Most are either harmless to other life, including humans, or are *vital* to the existence of life forms. For example:

- Bacteria in cows' stomachs are essential for digesting the cellulose in their food. Similarly, termites require a full complement of specific bacterial types in their intestinal tracts to digest the wood they consume.
- Organic waste substances in sewage and other organic materials are degraded by bacteria and transformed into material that plants depend on for growth.
- Certain species of bacterium cause cheese and other dairy products to be formed by metabolizing constituents of milk.
- Some bacteria add needed nitrogen to soil.
- The coliform bacteria living in our intestines are essential for a comfortable balance of digestion. If we go to a foreign country and ingest different bacterial types, we can suffer an imbalance in our digestive process.

Certain bacterial types are known to cause diseases. These pathogens can infect virtually every region of the human body. For example:

- Meningococcal bacteria infect the spinal and brain membranes.
- Diphtheria-causing bacteria have a unique affinity for the throat.
- Tubercle bacteria invade the lungs.

Bacteria, like all other organisms, continue to evolve according to the principle of natural selection. For example, the deadly lung disease tuberculosis, caused by the bacterium *Mycobacterium tuberculosis*, has reemerged as an unfortunate chapter in medical history due to the ongoing process of natural selection in this bacterial strain. By the middle of this century, it appeared that tuberculosis was essentially wiped out by treating infected individuals orally with a combination of two drugs, rifampin and isoniazid. Any person can be infected by tuberculosis, but those who have a normal immune system usually do not get the disease because their natural defense mechanisms are able to kill the bacteria or to keep the bacteria walled off in small scars in the lung, preventing the tiny organisms from growing and destroying lung tissue. If a person's immune system cannot keep the bacteria isolated and in check, they multiply in great numbers, thereby causing overwhelming infection and destruction of the lung tissue. The drugs mentioned above stop this disease process only when the program of treatment is followed faithfully and for many months.

In the mid-1980's there was a dramatic increase in the number of homeless people, those addicted to intravenous drugs, and sufferers of AIDS—all groups whose members typically have dangerously depressed immune systems. Since the *Mycobacterium* thrives on human hosts with such depressed immune systems, tuberculosis made a comeback in inner-city populations. The lifestyles of those groups of highly susceptible individuals are not conducive to a complete course of drug therapy to clear the infection. In other words, those most likely to be infected do not take their medicine properly or effectively, thereby providing a window of opportunity for the bacteria.

Some *Mycobacterium* organisms possess genes that render the drugs ineffective and have thereby allowed natural selection of those organisms to take place, resulting in new strains of antibiotic-resistant *Mycobacterium* that cannot be killed by any known drug. As a result, thousands of new cases of tuberculosis are being diagnosed every day as the newly adapted and infectious bacteria are spread by coughing and close contact, particularly in large cities worldwide. New drugs are currently being developed, but many people will die before the genetics of resistance is understood and effective therapeutic strategies take hold once more.

Certain bacteria can survive independent of other organisms, while others are parasites. *Viruses,* on the other hand, are never fully independent and are not cells. Each is a genetic system dependent on cells to survive. In other words, viruses need to grow inside animal, plant, or bacterial cells. They're much smaller and simpler than bacteria, consisting of only nucleic acid, a coat of protein, and in some cases fatty and carbohydrate material.

Figure 16-5

DRAWING OF TYPICAL VIRUSES

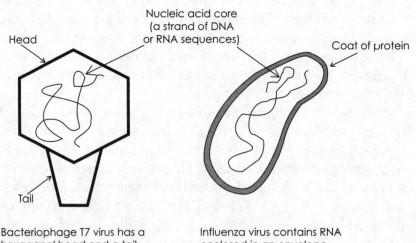

Head

Nucleic acid core
(a strand of DNA
or RNA sequences)

Coat of protein

Tail

Bacteriophage T7 virus has a hexagonal head and a tail structure which injects bacteria with the phage's DNA

Influenza virus contains RNA enclosed in an envelope

Some viruses called bacteriophages (or bacteria-eaters), like the one depicted in Figure 16-5, insert a spikelike device into bacteria to use the larger organism's enzymes and metabolic processes to reproduce. Though certain species of viruses and bacteria cause disease in humans and virtually all other species, obviously there is no harm "intended." They are simply living in the particular environment in which they evolved and adapted, reproducing and dividing like any other organism.

The Primordial Cell Evolves Into All of Life

Evidence in Slime Molds and Sponges

After one-celled organisms first appeared on Earth, millions of additional chemical and molecular changes developed under the pressure of natural selection over the next hundreds of millions of years, including the appearance of the DNA molecule, which gradually took over the self-replicating duties of RNA. Single-celled protozoans and other one-celled organisms obviously were enormously successful in evolutionary terms, and still are. But looking at humans and at all the other successful *multicelled creatures*, or metazoans, we see that it became advantageous for an organism to consist of more than one cell. About 600 to 800 million years ago, multicellular organisms first appeared. Through examination of the fossil record and today's creatures, the particular course of development from one-celled to multicelled life is fairly clear. We can see strong evidence of that evolutionary pathway in the organization of the "lower phyla," beginning with simple multicelled animals that evolved from the single-celled protozoans. For example, the organism we know as the slime mold exists at certain times of its life cycle as a single cell that can move around by means of a flagellum (long tail-like structure), like many protozoa and bacteria. At other times, many of these individuals gather together, their cell membranes coalesce, and a large mass containing thousands of nuclei moves about functioning as a single organism ingesting food. To replicate, this mass of protoplasm forms a platform with a stock that produces spores, much like a plant. This fascinating creature has features of both plant and animal cells, and scientists are not sure just how to classify it. There is little doubt that the cells that constitute the slime mold represent a very early life form that recapitulates much of what we understand about the first self-replicating cell on Earth.

As another example of early evolution from single to multicelled organisms, volvox proliferates in freshwater habitats about the globe. Volvox is simply a collection of individual single-celled protozoans living together in a colony in which they adhere to each other. Apparently this configuration helps the volvox survive and marks an early evolutionary swing toward the successful multicelled organism.

This pathway can be seen easily in the next most complex life forms that we know as the poriferans, or sponges. There are hundreds of different kinds of sponges, but each one consists of *three simple cell types* that are organized to carry out specific functions and that live together as a single organism: one cell type that moves water around an open body cavity, one that secretes a silica-based tissue for support, and one that digests small food particles drifting by. If the three cell types that make up a sponge are broken apart, they are capable of recognizing "self" and will stick together again in their own primitive way, but only to each other. However, these cells are not yet organized into tissues.

We need not travel very far up the evolutionary ladder before we find cells that make up tissues—that is, groups of interacting cells of similar structure and function, such as skin, muscle, and nerve. These can be found in the jellyfishes, anemones, and starfish. As simple as they are, these creatures can feel, and many are quite mobile. Starfish even have primitive cells organized to detect light at the tips of their arms.

The next steps up in evolutionary complexity are the flatworms, the roundworms, and then the segmented worms (see Figure 16-6). The arthropods (insects, crustaceans, etc.) followed several hundred million years later, and the amphibians, reptiles, and mammals were not far behind. The process of speciation and the evolutionary pressures of natural selection were slowly creating thousands of new, more complex species. Yet each cell of

Figure 16-6

FLATWORM AND EARTHWORM

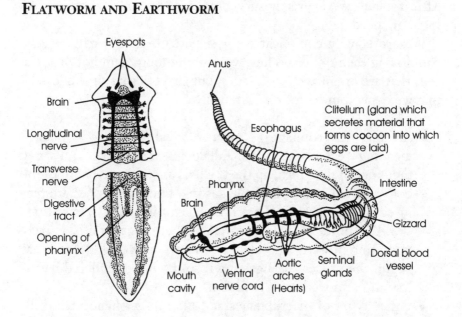

every organism functioned on the same principles of energy production and replication as those of the protozoans, slime mold, volvox, and sponges.

Knowledge of cell function and cell evolution constitutes a major aspect of evolution. It is also the basis for understanding reproduction and the origin of all life in the primordial soup.

Mental and Physical Ascent of Humans from a Single Cell Is Astounding

Macrophage Outposts and Human Cells That Crawl

Over a 70-year life span, the human brain holds about *100 trillion* bits of information. By comparison, the entire *Encyclopedia Britannica* contains only about 200 million bits. The superior power of the human brain is a result of natural selection, in the same manner as any other traits that confer a survival advantage. As discussed in Chapter 14, it is clear that our early ancestors who wandered the African plains were already surviving more by guile than by brute force or speed.

We are equally complicated physically. Of the 60 trillion cells comprising each of our bodies, we count among them not only the cells that are organized into tissues, but also millions of single cells that we depend upon for our survival:

- Alveolar macrophages ("macro" meaning "big," "phage" meaning "eater") consume inhaled dust particles and carry them out of the lungs, up the trachea, and ultimately out of the body. Two alveolar macrophages in a human lung (A) are pictured in Figure 16-7. The rounded, ruffled cell (B) is stationary. The other macrophage (C) in the foreground is actively migrating toward an inhaled foreign particle (D) and has begun the process of ingesting it so it will be rendered harmless and removed from the lung.
- Other types of macrophages and migrating phagocytes (cell-eaters) course through our blood vessels, gather dead blood cells, and engulf potentially infectious organisms.
- Other macrophages and blood cells fight cells that have turned cancerous.
- The more commonly known white blood cells (leukocytes), which we see as pus, ingest bacteria, dead tissue cells, protozoa, and other foreign bodies.

These are all part of the class of cells that are part of our body's defense mechanisms. As seen by the above examples, they engulf and destroy bacteria, viruses, and other foreign substances, as well as worn out and abnormal cells. They originate in the bone marrow and then venture out to assume their appointed duties at numerous outposts throughout the body. Although they exhibit many characteristics similar to bacteria and mites (like those living on your face) and similar to the free-living one-celled amoeba and protozoan, they are a product of *human* genes. In other words, our own DNA is programmed to create these "lively animalcules." They don't join us later, like bacteria, mites, amoebas, and protozoans.

Figure 16-7

TWO ALVEOLAR MACROPHAGES IN HUMAN LUNG—ONE IN FOREGROUND (C) REMOVING FOREIGN PARTICLE (D)

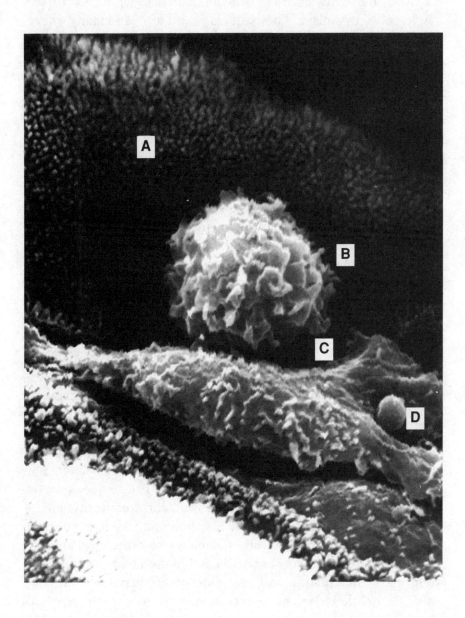

As stated earlier, the clear unity among life's diversity is seen in cells. Amazingly, if cells from any tissue or organ in the human body (skin, kidney, heart, lung, etc.) are cut away and placed in a dish with nutrients, each cell will immediately begin to act like a single-celled organism. They will crawl and feed, and many will reproduce by cell division. As each cell moves, it searches the space ahead of it and can pull back or bind to other cells, ultimately forming rudimentary tissue in the culture dish.

With 100 trillion bits of information in our brains and with a makeup and function that produces one-celled organisms that police our bodies as part of an elaborate defense system, we are incredibly complex beings, intellectually and physically. Even if one fully subscribes to the 65-million-year evolutionary process (from ancestral primates into *Homo sapiens*) described in Chapter 14, it might still be difficult to accept the idea that if you go back farther in time, we have actually evolved from a one-celled organism. How could that simple cell be the universal ancestor? Yet, in Part Five we saw that acceptance of great time spans was necessary to understand Darwin's theory. Within the same huge time span, the process of natural selection resulted in the evolution of one-celled organisms into multicelled life. Emotion and myth about human origins must give way to the facts we have in our grasp today.

Was the development of complex organisms such as humans inevitable, or did it occur by chance and historical accident? This question has been debated for decades and has led to many refinements of evolutionary theory, such as Stephen Jay Gould's theory of punctuated equilibrium, and the emerging theories about self-organization and the "science" of complexity. Entire books are devoted to the principles of self-organization and complexity, such as *At Home in the Universe* by Stuart Kauffman, in which he contends that there are general laws and principles of order that apply to complex biological systems and, in combination with natural selection, make the creation and evolution of life inevitable. In commenting on the Mars meteor that appears to contain traces of organic molecules, James Ferris of Rensselaer Polytechnic Institute says, "There is an inevitability to it . . . Life is a natural outcome of the physics, chemistry, and other laws of our universe."

The question of inevitability versus chance might never be answered conclusively. However, as we near the end of the century in which we came to understand the cell and its workings, we see that scientists have been able to determine physical processes that were responsible for self-replication and growth and that pushed the original one-celled organisms to evolve into complex plants, animals, and humans. Mankind has always been fascinated by its place and role in the universe and by beginnings such as this. The creation of the universe. The beginning of life. The first humans on Earth. Einstein once said that Copernicus taught us to be modest by showing that the Earth and mankind were not the center of the universe. Darwin said we should be proud for having ascended to the pinnacle of all forms of life, the only species that is self-conscious, contemplates its fate, and understands history. Now, scientists have been able to reconstruct the mix of elements that formed the first organic molecules, and trace evolutionary history from the first cells to 60 trillion cells working together. Whether this evolutionary path was inevitable or accidental, 20th-century biochemists and biologists have explained to us the essence of our very existence.

Chapter Seventeen

Beads on a Wire

"For Weismann, the germ-cells are immortal . . . Each generation hands to the next one the immortal stream unmodified by the experience of the body . . . The body is transient-temporary. Its chief 'purpose' is not its individual life, so much as its power to support and carry to the next point the all-important reproductive material . . . [But] the germ-plasm must sometimes change—otherwise there could be no evolution . . . Is there some internal, initial or driving impulse that has led to the process of evolution? . . . We can only reply the assumption of an internal force puts the problem beyond the field of scientific explanation."

THOMAS HUNT MORGAN
Heredity and Sex (1913)

The city of Brno sits at the confluence of the Svratka and Svitava rivers in what is now the Czech Republic. Founded in 1243, it survived the ravages of the Hussites, the Bohemians, the Swedes, the Silesian War, and Napoleon, and became the site of a now famous series of experiments that resulted in the first steps toward understanding how biological traits and characteristics are transmitted from parents to their offspring.

Mendel Formulates the Basic Principles of Genetics

Uncut Pages

The folklore concept that heredity is passed on through our blood is reflected in a number of common terms, such as "bloodlines," "half blood," "mixed blood," "bad blood," "blue blood," "royal blood," and "blood relative." Once again, we can look to Aristotle for the origin of this mistaken belief, yet it was accepted through the 19th century by most biologists, including Darwin. (Ironically, red blood cells are the only cells in our bodies that do not contain genes, since those cells have no nuclei.) Gregor Mendel (1822–1884) played a pivotal role in dispelling the old beliefs about inherited traits and establishing the study of heredity as a biological science. Born in Heinzendorf, Austria, Mendel was the son of impoverished peasant parents, and developed an interest in agriculture in his youth on the small family farm and orchard. He entered the Augustinian monastery in Brno in 1843 and was ordained a priest four years later. The Augustinians paid for and supported his continued interest in the natural sciences, sending him to the University of Vienna, where he studied physics, chemistry, mathematics, zoology, and botany. He returned to Brno in 1854 and began to teach natural science at the high school, where several colleagues were engaged in scientific pursuits and later founded the Natural Science Society. Mendel joined that group and was able to expand on his particular interest in botany through the many science books available at the school and monastery libraries.

Beginning in 1856, Mendel began experiments crossing varieties of pea plants in the small garden at the monastery to observe the statistical results of combining seven characteristics of the plants, including seed color, seed shape, tallness versus dwarfishness, and four other factors. For example, he wanted to see whether the offspring of a tall pea plant that was "mated" with a short pea plant would be tall, short, or medium height. Over the next several years of experiments in the garden and after analyzing literally thousands of pea plants, he determined the numerical ratios that came out of these experiments. From those ratios he established five principles that apply equally to all living things and endure to this day:

1. Each physical characteristic of a living organism (from one-celled protozoa to human beings) is the product of a specific "hereditary factor," which Mendel envisioned as some type of particle (now known as a gene).

2. These hereditary factors exist in living things in *pairs*. For example, the mother's pair might consist of a factor for green eyes and a factor for hazel eyes, and the father's pair, green and blue.

3. With respect to each such characteristic, only one of the two factors (e.g., for green eyes) that exist in the mother and one of the two (e.g., for blue eyes) in the father are passed to each of its offspring.

4. There is an equal probability that either one of the mother's factors and either one of the father's factors will be inherited by their offspring. So, under the example, the result could be any one of these combinations of hereditary factors for eye color in their child:

From Father/Mother
- Green/Green
- Green/Blue
- Hazel/Green
- Hazel/Blue

The offspring's eyes and all of its other physical characteristics are never a *mixture* of factors.

5. Some factors are dominant and some recessive. Therefore, if the factor inducing green eyes were dominant over the blue or hazel factors and if the blue factor were dominant over the hazel factor, 75 percent of the offspring of the two hypothetical parents would have green eyes (that is, the first three examples in number 4, above), and 25 percent of them would have blue eyes (the last example above). The only way for the child to have hazel eyes would be if both parents provide the recessive factor for hazel eye color.

In 1865 (six years after publication of Darwin's *The Origin of Species*) Mendel presented two papers (titled "Experiments with Plant Hybrids") to the Natural Science Society in Brno, explaining the results of his pea plant research. Yet, no one in this learned body recognized the application that Mendel's findings could have to understanding inheritance or evolution. Mendel even pointed out to them that no scientist had previously made it possible "to determine the number of different forms under which the offspring of hybrids appear, or . . . to ascertain their statistical relations." To the other members of the society, his work was mathematics, which they thought had little application to botany or natural philosophy. His paper was published in 1866 along with the other proceedings of the society for the previous year, and copies of the entire volume were routinely distributed to academic libraries in London, Paris, Vienna, Berlin, Rome, and other major cities throughout Europe, as well as in the United States. Mendel's paper went largely unread and unnoticed, except by the church authorities in Moravia who admonished him for what appeared to be Darwinian views, and reminded him of the church's stance on evolution. The last eighteen years of Mendel's life were uneventful from a scientific point of view. He returned to his monastic work, was elected abbot of his monastery in 1868, and was honored and respected for other work in the community. Meanwhile, his revolutionary paper lay dormant on obscure library shelves in those cities throughout the world.

Though Mendel collected all of Darwin's articles and books before and after *The Origin of Species*, there's no evidence that the two men ever corresponded with each other. Even if they had, it's difficult to say what might have come out of such contact, but it probably would not have altered the course of the history of heredity as a biological science, particularly because specific evidence of Mendel's hereditary "factors" had to await the work in cell research discussed in Chapter 16 that led to the birth of genetics itself in the twentieth century.

Other chapters in the story begun by Darwin with *The Origin of Species* were slowly and quietly being written in the latter part of the nineteenth century by Walther Flemming and August Weismann. As mentioned in Chapter 16, Flemming's highly accurate description of cell division in his 1882 book *Cell-Substance, Nucleus, and Cell Division* showed that cell reproduction involves the transfer of chromosomes from the parent to "daughter" cells through mitosis. August Weismann (1834–1914), a German biologist, was familiar with Flemming's findings and recognized they did not answer his questions about the division of the unique cells known as egg and sperm. With Mendel's experiments unknown to Weismann and the rest of the world, Weismann wrote that the sperm and egg cells of animals must contain "something essential for the species, something which must be carefully preserved and passed on from one generation to another." This was his "germ-plasm" theory, first published in *The Germ-Plasm: A Theory of Heredity*, in 1886, which concluded that all living things contain a special hereditary substance. He also recognized that if *all* the hereditary substances from the two parents became mixed together in the fertilized egg, there would be a geometric increase of those substances, which eventually would be impossible for the cells to store. Thus, as part of his theory, Weismann predicted that the material of heredity is carried by Flemming's chromosomes and that there is a unique form of cell division (now called "meiosis") resulting in each "daughter" nucleus receiving only *half* of the paired chromosomes from each of the parent cells' nuclei. Applied to humans, for example, Weismann's theory would hold that each egg cell and each sperm cell contains 23 chromosomes instead of the 23 *pairs* contained in every other cell in the human body (except red blood cells, which have no nucleus, thus no chromosomes).

As the 19th century came to a close, a number of biologists were conducting research to test Weismann's theory and gain insight into the behavior of the sperm and egg cells during this process. Some of that research involved experiments much like Mendel's of thirty-five years earlier. It was at this point—on the threshold of the century that was to become the most incredible in all of scientific history—that Mendel's classic paper was rediscovered, retrieved from its scholarly grave in March 1900 by Hugo DeVries of the Netherlands through the publication of his two papers on the hybridization of plants. In one of his papers, DeVries noted that Mendel's study "is so rarely quoted that I myself did not become acquainted with it until I had concluded most of my experiments, and had independently deduced the above propositions." When other scientists went to the library shelves to read their copy of the 1865 *Proceedings of the Brno Society for the Study of Natural Science*, they found that the pages of Mendel's paper had not been cut. That is, no one had separated the edges after the volumes were sent to the various libraries. *In those volumes, the paper had never been read.* For thirty-four years it had not seen the light of day, literally and figuratively, but the words on those pages were soon to be transformed from obscure musings into the basic laws of heredity.

The Science of Genetics Is Born

Mapping Chromosomes

With Mendel's ideas introduced into the mainstream of research on heredity, the significance of Flemming's mitosis and Weismann's meiosis became apparent. In a 1902 paper in the *Biological Bulletin*, later refined in an article titled "The Chromosomes in Heredity," the American geneticist Walter S. Sutton (1877–1916) provided the first conclusive evidence that *chromosomes carry the units of inheritance and that they occur in distinct pairs*, as Weismann had predicted. The evidence was obtained through Sutton's experiments on sperm formation in grasshoppers, in which he observed that each chromosome in the cell nucleus becomes paired with a physically similar chromosome. As shown in Figure 17-1, the paired chromosomes then contract, duplicate, and later separate, with each going to a different sperm or egg cell. The physical pairing of the paternal and maternal copy of each chromosome is unique to meiosis. Contrary to mitosis, wherein the paternal and maternal chromosomes simply line up with each other before anaphase and separation (see Figure 16-3), in meiosis the chromosomes replicate and then bond. In mitosis, each cell produces two new cells, each with a full complement of paired chromosomes, whereas the process of meiosis involves *two separate cell divisions*, resulting in *four* new cells, each with a set of half the number of chromosomes of other cells. For purposes of illustration, Figure 17-1 shows only two chromosomes in the egg or sperm cell's nucleus. However, the nucleus of every human egg or sperm cell actually contains 23 chromosomes. Human egg cells are large enough to be barely visible to the naked eye.

Figure 17-1

Stages of Meiosis

Meiosis is the Unique Process of Cell Division for Egg Cells and Sperm Cells

1. Paternal (black) and maternal (white) chromosomes contract to form in cell nucleus

2. Prophase—Replication of genetic material occurs, followed by pairing of chromosomes with the same kinds of genes, which then bond and exchange genetic material. This exchange is called recombination. (See Figure 17-2)

3. Metaphase— Paired chromosomes line up at "equator"

4. Anaphase I— Chromosomes separate in first cell division

5. Telophase I—Two new cells separate and nuclear envelope reforms

6. Anaphase II— Chromosomes separate in second cell division

7. Telophase II— Resulting in four cells, each with half the number of chromosomes (23) as in mitosis, and each with genetic material from both parents as a result of recombination

Meiosis also differs from mitosis in that the "contents" of the chromosomes (the genes) in these egg and sperm cells are solely responsible for the hereditary makeup of the offspring. Thus, while mitosis relates to the everyday life of a variety of cell types, meiosis involves the fundamental processes of genetics and evolution. Sutton's experiments on meiosis converted Mendel's abstract mathematics and Weismann's predictions into physical reality by identifying the "collections" of these particles (later called "genes") on chromosomes as the hereditary factors.

Meiosis and the above discussion concerns division of egg and sperm cells. It is not about fertilization of the egg cell or reproduction of the entire organism (such as a human). However, once meiosis was understood, it became clear that *the essence of heredity is the duplication of the genes, the carriers of genetic information.* Beginning at the moment the egg cell is fertilized, the process of mitosis discussed in Chapter 16 (see Figure 16-3) results in this one fertilized egg cell ultimately dividing into trillions of cells. "Directed" by the genes, these cells differentiate into hundreds of types, each with a specific function (liver cells, nerve cells, skin cells, brain cells, etc.). As we will explore in the next chapter, it would still be decades after the work of Sutton and other geneticists before scientists were able to figure out the specific chemical makeup of genes.

Beginning in 1903 and continuing for decades, the following scientists expanded on the discoveries of Darwin, Mendel, Flemming, Weismann, and Sutton, and refined our understanding of the principles at work when offspring inherit their parents' genetic makeup:

- Herman Nilsson-Ehle (1873–1949): Beginning in 1900 and continuing to his retirement in 1939, this Swedish geneticist conducted research with wheat strains and other plants, refining and confirming Mendel's five principles of inheritance. Over the course of his career he opened up new fields of research in genes and chromosomes, and advanced the knowledge of mutations.
- Edward M. East (1879–1938): His pioneering work in plant genetics and botany beginning in 1900 included experiments with tobacco plants in which he accurately concluded that

spontaneous mutations in the genes themselves accounted for certain changes throughout generations of such plants without any change in environmental conditions. These mutations can be essential in the process of natural selection, because when such a mutation confers an advantageous trait, it is selected by an increased chance of that organism surviving. That mutated trait is then passed on to the offspring.

■ Thomas Hunt Morgan (1866–1945): Through his experiments with the fruit fly *Drosophila* (which produces a new generation in two weeks), Morgan, a geneticist and zoologist, made the monumental discovery that chromosomes are *not permanent structures*. In 1909, he adopted the word "gene" (first used that year by the Danish botanist and geneticist Wilhelm Ludvig Johannsen) to refer to one of Mendel's "hereditary factors." Over the next several years, in collaboration with three of his graduate students at Columbia, Morgan not only confirmed Sutton's theory that each chromosome carried a collection of genes "strung out like beads on a wire," but he discovered that the position of each of those beads could be "mapped" and identified within precise regions of the chromosomes.

More important, Morgan and his group were the first to prove that during the stage when chromosomes become paired and then contract, they may *exchange* genetic material between chromosomes of maternal and paternal origin, as noted in the prophase stage of meiosis (see Figure 17-1). This is called crossing over. When the "recombined" chromosomes separate into newly formed egg or sperm cells, those new cells contain a unique array of genes in their chromosomes as a result of that exchange.

The earlier figure on meiosis (Figure 17-1) illustrated this exchange of genes between chromosomes. Figure 17-2 provides greater detail, depicting the actual strands of genetic material (now known to be DNA) that make up the chromosomes. Recombination of genetic material also occurs when single-celled bacteria or simple fungi reproduce by cellular division. However, the term generally refers to the exchange of DNA sequences

Figure 17-2

GENETIC RECOMBINATION DURING MEIOSIS

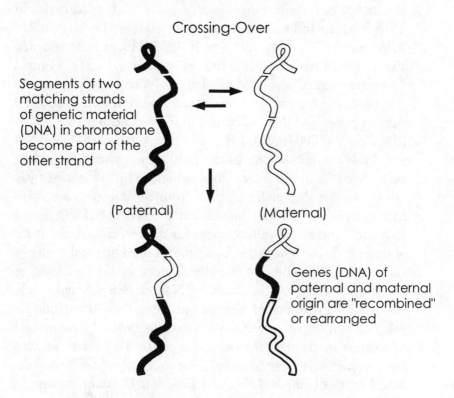

Crossing-Over

Segments of two
matching strands
of genetic material
(DNA) in chromosome
become part of the
other strand

(Paternal)   (Maternal)

Genes (DNA) of
paternal and maternal
origin are "recombined"
or rearranged

(i.e., genes) between the paternal and maternal chromosomes during the normal process of meiosis in humans and animals.

The critical significance of recombination is that the egg and sperm cells are, of course, the only cells that determine the complete genetic makeup of the offspring. The *recombined* genetic material is passed on to subsequent generations. Morgan and his colleagues proved that the process of variation, which circumstantially explains evolution, is *not* due to significant mutations occurring in every new generation, but to recombination of the "beads on a wire"—the genes. Thus, Morgan established a firm relationship between Darwin and

Mendel, and discovered that Mendel's factors have a physical basis in chromosome structure. He also laid the foundation for the theory of linear arrangement of genes in chromosomes. In 1915 Morgan and his assistants published these findings in *The Mechanism of Mendelian Heredity*. In 1933, Morgan was awarded the Nobel Prize for his work in genetics. In the Human Genome Project, discussed in Chapter 19, scientists are now in the process of identifying the exact location of all 100,000 human genes and their relationship to one another.

- R.A. Fisher (1890–1962), J.B.S. Haldane (1892–1964), and Sewall Wright (1889–1988): In the 1920's, these three geneticists, well versed in mathematics, independently but simultaneously calculated that the small variations from chromosome recombinations as well as the spontaneous mutations deduced by Edward East could mathematically account for the vast changes in living organisms through the periods of time inferred by fossil evidence and required for evolution by natural selection. Six decades after the Brno Society for the Study of Natural Science had sorely missed the importance of Mendel's statistics, these three individuals founded the subject of population genetics and gave natural selection a mathematical basis and explanation. Ronald Fisher's book in particular, *The Genetical Theory of Natural Selection*, published in 1930, showed that the slow but constant change in genes and chromosomes explains Darwinian evolution.

At age seven Sewall Wright wrote a booklet titled "The Wonders of Nature," and later became fascinated by genetics when he read about Mendel in the *Encyclopedia Britannica*. His first paper was published in 1912 (when he was a graduate student, mentored by Edward East) and his last in 1988. By combining his interest in genetics with his formidable talent in math, which he taught himself, Wright performed pioneering work in mathematical population genetics and evolutionary theory, including his treatise on those subjects, *Evolution and the Genetics of Populations* (published in four volumes, 1968, 1969, 1977, and 1978). Since he was the sole survivor of those who established genetics in the early 20th century, Sewall Wright's death in 1988 marked the end of an era.

- Barbara McClintock (1902–1992): After receiving her Ph.D. in botany from Cornell University in 1927, this American geneticist taught botany and genetics for several years and then took a research position at the Cold Spring Harbor Laboratory in New York where she performed a series of experiments on the colors of corn kernels that provided new and definitive information about recombination, the reality and characteristics of linkage groups of genes, and the relationship among particular genes. She remained at Cold Spring Harbor for fifty years and won the Nobel Prize in 1983 for her pioneering work done in the 1940's and 1950's on genetic function and organization.

Once the gene had been conclusively identified as the essential "heredity factor" postulated by Mendel, the focus turned from hereditary principles and concepts to the physical and chemical components of this mysterious and extraordinary bit of matter in the cell's nucleus. By 1932, James Chadwick had discovered the neutron, the third subatomic element in the *atom's* nucleus, and ten years later physicists learned to harness the immense power of the atom. Curiously, however, it would take many more years before we fully understood the structure and inner workings of the cell's nucleus, even though it is much larger since it consists of *combinations* of atoms in the form of the crucial and highly complex organic molecules that control all life. Yet it was that very combination and complexity that caused the mystery to persist. Amazingly, in the early years of the 20th century, at almost precisely the same time that the famous physicist Ernest Rutherford predicted that the forces of the atom's nucleus might never be fully understood, geneticist Thomas Hunt Morgan said of genes in the cell's nucleus that "an internal force puts the problem beyond the field of scientific explanation." A generation later, in the seventh and most recent great scientific discovery—which will be explored in the following pages—this "internal force" was identified and its structure and function became understood, thus closing the circle of knowledge on the three interrelated and basic aspects of all life: *its origin, reproduction, and evolution.*

Part Seven

The Structure of the DNA Molecule

In Focus

The last of the seven greatest scientific discoveries in history is the configuration of deoxyribonucleic acid, commonly known as the DNA molecule. In turning from Part Six to Part Seven, we are advancing from cell structure and principles of genetics to the physical components and inner workings of certain molecules in the cell's nucleus, including DNA. Since all aspects of living things are ultimately determined at the molecular level, it was necessary to gain an understanding of the molecule of life to understand life itself. In Part Seven we will see how we came to that understanding. We will see the structure and function of the organic molecule that is responsible for the existence of all life and is the mechanism by which evolution takes place. We'll recount the events that led two men to discover that structure in February 1953, less than a century after publication of *The Origin of Species.*

Genetics brought together diverse disciplines, such as biochemistry, biophysics, and microbiology. Today, this combination is called molecular genetics—the study of the molecular structure of genes and the way genes control cells. In this all-encompassing field, we have now amassed an extraordinary amount of knowledge about the process that creates the molecules of life and the process by which those molecules direct the formation and creation of the rest of the organism. In fact, we have gained so much knowledge of how we are made that work is in progress to map the entire human genetic code and apply this knowledge to cure genetic diseases. We will examine this amazing revolution and the enormous potential it holds for our lives.

Chapter Eighteen

The Backbone of Life

"No new scientific laws are required to explain living organisms' complexity and organization . . . The essence of life . . . is the ability of living things to extract energy from their environment and to use that energy both to build up their own complex structures and to copy themselves."

JOHN GRIBBIN
In Search of the Double Helix (1985)

". . . The book of life is very rich; a typical chromosomal DNA molecule in a human being is composed of about five billion pairs of nucleotides. The genetic instructions of all other life on Earth are written in the same language with the same code book . . . This shared genetic language is one line of evidence that all the organisms on Earth are descended from a single ancestor, a single instance of the origin of life some four billion years ago."

CARL SAGAN
The Dragons of Eden (1977)

Exactly what are genes and what are they made of? In search of the answer to this question, scientists first explored the chemical and molecular composition of genetic material, then its most basic atomic structure. As Albert Lehninger stated in *Principles of Biochemistry*, "Living things are composed of lifeless molecules [that] conform to all the physical and chemical laws that describe the behavior of inanimate matter." Yet the molecules that provide the essence of life have a unique and extraordinarily complex structure. Understanding the complexity of this structure, including the particular functions of the molecules and how they sustain life through reproduction, became a fascinating and elusive goal of scientists in the 20th century.

Identification of Nucleic Acids and the Right Molecule Launch Molecular Genetics

Suggestion of a Secret Code

Decades before geneticists such as Morgan and McClintock expanded our knowledge of the organization of genes and chromosomes, a number of chemists were exploring the makeup of the cell's nucleus. They did not realize that their work would ultimately provide key information about heredity. In 1869 the Swiss biochemist Friedrich Mieschner (1844–1895) first proposed that all cells' nuclei probably have a specific chemistry. In succeeding years he discovered various substances in the nucleus, which he separated into *protein and acid molecules*—hence, the term "nucleic acids."

A Russian-born chemist, Phoebus A. T. Levene (1869–1940), was also a pioneer in the study of nucleic acids. After receiving his M.D. from the St. Petersburg Imperial Medical Academy in 1891, Levene fled anti-Semitism in Russia and settled in New York City, where he studied chemistry at Columbia University and then engaged in chemical research from 1905 to 1939 at the Rockefeller Institute for Chemical Research. In 1909 he correctly identified ribose as the sugar in one of the two types of nucleic acids, ribonucleic acid (RNA), and he identified certain components in the other nucleic acid, deoxyribonucleic acid (DNA). With nucleic acids and protein in the nucleus, Levene and many of his colleagues were convinced that the complex and plentiful *protein molecules* (and not DNA) stored all the genetic information in the chromosomes. Unfortunately, researchers labored under this false assumption for decades as they pursued the source of some type of gene-copying mechanism in the proteins. Levene's theory of the purpose of

DNA—that it merely holds protein molecules together—turned out to be incorrect.

The work that led to correcting that false assumption was begun in 1928 by Frederick Griffith (1881–1941), an English bacteriologist. His research on pneumonia-causing bacteria (pneumococci) led to his discovery that a certain unknown substance from the cells of one strain of *dead* pneumococci was able to enter a different and living strain and cause the live strain to pass the dead strain's hereditary characteristics to the live strain's offspring. Another bacteriologist, Oswald T. Avery (1877–1955), together with his colleagues, became aware of the significance of Griffith's work and spent ten years trying to identify the agent that was the essence of genetic transformation in the bacteria. Finally, in 1944 Avery and his collaborators published the results of their lengthy research, which clearly showed that it was DNA, not protein or RNA, that allowed the transport of hereditary information. This work launched the science of molecular genetics.

Subsequent research in the United States by the Austrian-born biochemist Erwin Chargaff (1905–) determined the proportions of the four compounds present in DNA: adenine (A), cytosine (C), guanine (G), and thymine (T). The term in chemistry for these substances is "base," meaning something that reacts with acids to form salts and has certain other chemical characteristics. In 1950, Chargaff determined the exact proportionate amounts of the DNA bases within each molecule: *Guanine equals cytosine and adenine equals thymine*. Therefore, the amount of guanine and adenine *combined* equals that of cytosine and thymine *combined*. These are known as the Chargaff Ratios, and were to become a key element in the discovery of the structure of the DNA molecule.

A series of experiments in the 1940's and early 1950's by the American biologist Alfred D. Hershey (1908–), dealing with DNA of bacteriophages—viruses that infect bacteria—further confirmed the Avery group's conclusion that DNA, not protein, is the genetic material. In 1969 Hershey won the Nobel Prize for this work. Thus, the work by the early geneticists, as well as Avery, Chargaff, and Hershey, suggested that there was some sort of genetic information or code configured in DNA and passed on to offspring of all living things.

As mentioned above, nucleic acids come in two types: ribonucleic acid (RNA) and deoxyribonucleic acid (DNA). The bases are the same in both molecules, except that uracil replaces thymine in RNA. The chemical composition of DNA can be pictured like this:

Figure 18-1

BACKBONE AND BASES OF DNA

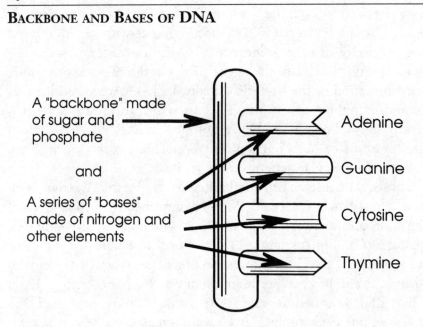

A "backbone" made of sugar and phosphate

and

A series of "bases" made of nitrogen and other elements

Adenine

Guanine

Cytosine

Thymine

As we will soon see, the *bases* of DNA and RNA are of critical importance in understanding what a gene is.

Genetics and Chemistry Combine with Physics to Bring Us to the Threshold of Discovery

Quantum Physics Revisited

The foundation of the next phase of this story dates back to the time of Röntgen's discovery of X-rays in 1895, and the work of

Thomson, Becquerel, Rutherford, Bohr, Einstein, Planck, and others. As discussed in Part Two, these scientists discovered the present model of the positively charged atomic nucleus (protons and neutrons) with negatively charged electrons orbiting the nucleus within distinct shells. Aggregates of atoms form into molecules by sharing electrons in the outermost shell or orbit of each atom—the outermost shell "needs" the additional electrons, resulting in the chemical bond between elements. Bonds between hydrogen atoms are slightly different than other chemical bonds in that they do not actually share electrons, but are brought together by the electrical attraction of the atoms. The same principles of bonding that apply to simple molecules like H_2O (water) also apply to the much more complex organic molecules we're dealing with in biology, which contain billions of atoms.

Those basic principles of chemistry and physics eventually combined with biology to give us the answer to the question asked earlier: *What is a gene?* In 1935, the German-born physicist Max Delbruck (1906–1981) coauthored a now famous paper with two biologists, which addressed the question of why short-wave radiation causes mutations in genetic material. Delbruck explained that the answer is found in the branch of physics called quantum theory, because mutation in living organisms involves molecules in the cell's nucleus being pushed over an energy barrier (measured in "quanta") from one stable configuration to a different stable configuration. In other words, he demonstrated that the physics of the atom itself determines the most basic biological processes. Influenced by Delbruck's paper, in 1944 Erwin Schrödinger (1887–1961), the famous Austrian theoretical physicist, wrote the book *What Is Life?* in which he further illustrated how quantum physics explains the stability of the molecules in the genetic structure. At that time the results of Oswald Avery's work on DNA had just been published, so Schrödinger's book did not state that genetic information was contained within DNA. However, Schrödinger's book had a great impact on several other physicists working in biology who later recognized the connection between DNA and Schrödinger's calculations.

Other physicists began to recognize the importance of Delbruck's and Schrödinger's groundbreaking work. In addition, the biochemists Linus Pauling (1901–1994) and J.D. Bernal (1901–1971) were making significant discoveries about the structure of proteins. Influenced by these developments, after World War II ended in 1945 a number of physicists turned away from fundamental physics to the exciting and burgeoning field of biology, where they saw a major role for physics in solving biological issues. They brought with them the thought processes characteristic of their field, that is, the concept that all material bodies (such as organic molecules) are arrangements of large numbers of atoms, and the behavior and property of these bodies can be explained by determining how atoms are joined together to form molecules.

For a physicist to make this transition to biology was like a doctor becoming an engineer—the fields seemed to have nothing in common. It was a bold decision. The vast majority of physicists and biologists didn't recognize the potential benefits in developing such an alliance. However, this handful of researchers in both fields began to engage in molecular biology, and the term "biophysics" was coined. In its infant stages, biophysics was merely a technique to apply the knowledge about atoms and molecules to biological issues, but it soon became an important part of molecular genetics.

One person influenced by Schrödinger and his work was Maurice H.F. Wilkins (1916–), a New Zealand–born British biophysicist who used X-ray diffraction studies to examine the molecular structure of DNA. X-ray diffraction, also called X-ray crystallography, is a phenomenon in which the regularly arranged atoms of matter (particularly a crystal structure) diffract, or scatter, the X-rays, which can then be recorded on a photographic plate, much like Wilhelm Röntgen's original discovery of X-rays in 1895. The pattern that shows up on the photograph provides important clues to the configuration of the atoms in the material, which would otherwise not be detectable because until very recently atoms were too small to be seen with any type of microscope. Wilkins first gained knowledge of X-ray diffraction through his use of the mass spectrograph in his two years of work on the Manhattan Project during World War II. After his return to England, he and a research student named Rosalind Franklin (1920–1958) used X-ray diffraction to take pictures of fibers composed of pure DNA. These pictures became a critical link in the chain of events that ultimately led to the discovery of the structure of the DNA molecule.

By the late 1940's, molecular biologists knew that DNA was the molecule of life, but they did not know its exact structure, and, therefore, how it carried out its functions or how it duplicated itself. However, the time was ripe for the work in genetics to converge with the new knowledge of the chemistry of the cell's nucleus. This convergence would reveal the molecule's secret. By applying the newly found information from biophysics to these other bodies of knowledge, two men fiercely pursued DNA in the hope they could entice it to whisper its secret into their ears before anyone else discovered it. This fateful and serendipitous pursuit resulted in one of science's and all of history's most magnificent moments.

Crick and Watson Meet and Begin Their Collaboration

Basing a Career on the Gossip Test

Francis Harry Compton Crick (1916–) was born in Northampton, England, the older son of a middle-class couple. His father, Harry Crick, and his uncle ran a boot and shoe factory founded by their father in Northampton, a town dominated by the manufacturing of footwear. At an early age, he showed an interest in scientific subjects, so much so that his parents bought him a set of Arthur Mee's *Children's Encyclopedia* when he was eight, since they were unable to answer his constant questions about the universe. "I read it all avidly," Crick later wrote. "It was the science that appealed to me most. What was the universe like? What were atoms? How did things grow?" Yet, as he marveled over the fascinating facts in the encyclopedia, he was struck by one great fear: "By the time I grew up . . . everything would have been discovered . . ." When he confided in his mother about this fear, neither knew how prophetic her reassurance would be. "There will be plenty left for you to find out," she told her son. By his own description, Francis Crick was not exceptionally precocious or outstanding in school, but his curiosity persisted. This desire to know the answers to the great questions he'd posed as a child, together with an unusual reasoning ability and affinity for logical thinking, were to chart his destiny.

After graduating from University College in London in 1937 with a degree in physics, he engaged in postgraduate research in hydrodynamics. This provided him with a great appreciation of the scientific method and with the idea that a variety of scientific disciplines could be as successful as physics had been in the early part of the 20th century. During World War II, Crick worked in the Admiralty Research Laboratory, designing magnetic and acoustic mines. After the war, at age 30, he was faced with the decision of what to do with the rest of his life. As he was to write later, he concluded that "what you are really interested in is what you gossip about." He realized he was not gossiping about physics, but about recent advances in biology. Using this unscientific methodology to determine his future, and considering his desire to engage in fundamental research, Crick narrowed his interests to "the borderline between the living and the nonliving." He was referring to the fact that no one had yet fully explained how lifeless atoms could create life or how organisms pass on their characteristics. Convinced that the gossip test had led him to his true calling, he became one of the handful of adventurous postwar physicists to cross over to biology.

In 1946 Crick took a job with the Strangeways Laboratory to study the physical properties of the cytoplasm of cells. After about two years there, he came to the realization that "one needed to discover the molecular structure of genes" to understand life and that "the most useful thing a gene could do would be to direct the synthesis of a protein, probably by means of an RNA intermediate . . ." At that time, Crick learned that a research unit dedicated to studying the structure of proteins using X-ray diffraction was about to be established at the Cavendish Laboratories of Cambridge University, the site of many earlier great discoveries in physics. In 1949, at the age of 33, Francis Crick transferred to this lab to become a graduate student again and learn about the three-dimensional structure of proteins, with the hope that this great laboratory's fame in physics could be extended into biology.

Meanwhile, James D. Watson (1928–) had obtained his doctorate from Indiana University at the tender age of 22 and accepted a

U.S. government fellowship to work under the biochemist Herman Kalckar in Copenhagen. By his own admission, he possessed the grandiose and vain notion that he would discover the secret of life. "My interest in DNA had grown out of a desire, first picked up while a senior in college, to learn what the gene was . . . It was certainly better to imagine myself becoming famous than maturing into a stifled academic who had never risked a thought." At the forefront of the research on X-ray diffraction of DNA were Maurice Wilkins, the New Zealand-born biophysicist mentioned above, and Wilkins's assistant Rosalind Franklin, at the newly created biophysics department at King's College in London. After attending a lecture by Maurice Wilkins in the spring of 1951, Watson "began worrying about where I could learn how to solve X-ray diffraction pictures." Without approval from the Fellowship Office in Washington, D.C., he left his fellowship appointment to take a job working on X-ray crystallography at Cavendish, where Crick had been since 1949.

Cavendish Laboratories was under the direction of physicist William Lawrence Bragg (1890–1971), who'd won the Nobel Prize in 1915 for his own X-ray diffraction work. This included the formulation of Bragg's law, an equation that *correlates the wavelengths of X-rays and the distance between "planes" of atoms.* At Cavendish, Watson was supposed to work on determining the complex structure of the protein myoglobin, a molecule in the muscle. However, like Wilkins and Crick, James Watson hoped that X-ray diffraction would help reveal the particular configuration among atoms in DNA.

According to Robert Shapiro, in *The Human Blueprint*, Watson arrived in Cambridge "as a socially awkward, untidily dressed, nervous, ambitious, and intense young American Ph.D." In the fall of 1951, Francis Crick met this brilliant and brazen young researcher. They soon learned they shared a passion for the elusive macromolecule, DNA, both being convinced it was the secret of life. Thus began the union that was soon to lead to the seventh great scientific discovery in history. Crick later wrote, "Jim and I hit it off immediately, partly because our interests were astonishingly similar and partly, I suspect, because a certain youthful arrogance, a ruthlessness, and an impatience with sloppy thinking came naturally to both of us. Jim was distinctly more outspoken than I was, but our thought processes were fairly similar." Ironically, neither Crick nor Watson was officially working on DNA at Cavendish. "I was trying to write a thesis on the X-ray diffraction of polypeptides and proteins," Crick explained, "while Jim had ostensibly come to Cambridge to . . . crystallize myoglobin . . ."

The Search Goes Under Cover, Then Becomes a Race to Beat the Americans

The Pauling Factor

At first, Crick told Watson that solving the X-ray diffraction patterns of the DNA fibers should be left up to Maurice Wilkins and Rosalind Franklin. However, they soon became impatient with this pair's slow progress, which was largely the result of Wilkins's and Franklin's dislike for each other. Thus, Crick and Watson undertook analysis of the patterns on their own. However, Crick did not get along well with his boss, William Lawrence Bragg. Also, Wilkins's and Franklin's director at King's College convinced Bragg that Crick and Watson were duplicating the King's College work. So in the fall of 1951, soon after Crick and Watson began their joint work, Bragg ordered them to cease their unofficial quest for the secret of life and return full-time to the mundane research they were supposed to be pursuing at Cavendish—proteins and myoglobin. Without rank and without any concrete results with which to dissuade Bragg, their only course of action was to pursue their obsession in secret. As Watson later explained in his book, *The Double Helix*, they could not even chance a casual question to Maurice Wilkins, since that "would provoke the suspicion that we were at it again." At first, they limited their work to discreet conversations at lunch. Soon, however, as Watson explained, he began to spend "the dark and chilly days to learn more theoretical chemistry or to leaf through journals, hoping that possibly there existed a forgotten clue to DNA."

But in January 1952, even this remnant of the secret DNA project was threatened with total dissolution when the Fellowship Office in Washington, D.C., revoked Watson's stipend and award, because he'd left Copenhagen for Cavendish, and demanded his immediate return to the United States. Not to be distracted from his stubborn pursuit, Watson defiantly wrote back that he found Cambridge intellectually exciting and that he had no immediate plans to return to the States.

Spurring the pair on with more intensity were the stories coming out of the California Institute of Technology, where the famous biochemist and eventual three-time Nobel Prize winner Linus Pauling appeared to be close to discovering the structure of DNA. Linus's son, Peter Pauling, was a research student at Cavendish while Watson and Crick were there, and exacerbated their fears with frequent reports about his father's progress on DNA. But late one night in June 1952, Watson developed a negative of an X-ray he'd taken of DNA material. "The moment I held the still-wet negative against the light box," said Watson, "I knew we had it. The telltale helical markings were unmistakable." Crick, the physicist and the expert in X-ray diffraction of macromolecules, spotted the crucial reflection in "less than ten seconds," thus confirming Watson's earlier suspicion that DNA had a helical, not a linear, configuration.

Nevertheless, by the end of 1952 they "remained stuck at the same place we were twelve months before" because they still didn't have proof of how many strands were in the helix or the knowledge of how to explain its structure from a chemical and molecular standpoint. That is, they couldn't explain how or why the components of DNA (the backbone and the four bases—adenine, guanine, cytosine, and thymine) all fit together. Thus, when Peter Pauling came into the lab one day in December 1952 and announced that his father had determined the structure of DNA, their hearts sank. They feared the race had been lost and the Nobel had slipped away. But as weeks went by without any real news coming from Pauling and Cal Tech, Watson and Crick thought unlikely that Pauling could have discovered the actual DNA structure, particularly because he didn't have access to the Wilkins/Franklin X-ray work at King's College.

Finally, Peter showed his father's manuscript to Crick and Watson, and to their immense relief it showed a *three-chain* helix with a sugar-phosphate backbone in the *center*. They had played with the idea of a three-chain molecule over a year earlier and were convinced it was not accurate. "Everything I knew," Watson recalled, "about nucleic-acid chemistry indicated that phosphate groups never contained bound hydrogen atoms . . . Yet somehow Linus, unquestionably the world's most astute chemist, had come to the opposite conclusion . . . We were still in the game."

However, they knew that when Pauling became aware of his mistake, he would pursue this work at an even faster pace to find the right structure. So, for the first time since Bragg had banned them from DNA, Crick and Watson openly pleaded their case to him—it was now imperative that they get back into the DNA business with both feet to beat out the scientists on the other side of the Atlantic. Crick and Watson estimated they had six weeks before Pauling would realize the errors he'd made in the three-chain theory.

Wilkins's X-Rays and Quantum Mechanics Lead to the Final Answer

No Experiments and a Wood and Metal Model

With at least a temporary reprieve, the first order of business was to find out the latest results of Maurice Wilkins's and Rosalind Franklin's work at King's College. When Watson walked into Wilkins's office in January 1953, he saw to his amazement that Wilkins and Franklin had the clearest evidence of not only a helical structure, but vital parameters that could enable Watson and Crick to fix the number of chains in the helix at two. Though these X-rays were a year old, Wilkins and Franklin failed to recognize their significance and were unable to decipher the helical aspects that were so clear to Watson and Crick. As Watson explained, "The instant I saw the picture my mouth fell open and my pulse began to race."

Watson raced back to Cavendish to begin building a two-chain model. "Important biological objects come in pairs," he observed. After hearing what Watson had learned at King's College, Bragg began to cheer him on in the race against Pauling, while Crick continued to tread more lightly, since he was not fully back in Bragg's good graces. The two proceeded up numerous blind alleys over the next several weeks until the American crystallographer Jerry Donahue convinced them of a particular quantum mechanical argument about hydrogen bonding and the pairing of adenine only with thymine and guanine only with cytosine (based on the Chargaff Ratios) that sent them in the right theoretical direction. In late February 1953, the day after they spoke with Donahue, Watson suddenly realized that "an adenine-thymine pair held together by two hydrogen bonds was identical in shape to a guanine-cytosine pair." This bonding and pairing solved the last remaining problem they'd faced, because it accounted for the two types of base pairs being identical in shape so the double helix structure would be made up of uniform angles and rotation. This was the crucial revelation that enabled them to begin making the final mathematical calculations and begin constructing a wood and metal model.

Incredibly, during their eighteen-month collaboration (from fall 1951 to March 1953) Crick and Watson never performed experiments to determine the structure of the DNA molecule. They performed limited research in areas related to their pursuit of the double helix such as chemical base pairing and taking X-ray photographs of the tobacco mosaic virus, but no experiments on the structure of DNA. They only engaged in theoretical discussions on the subject and built the models. Reflecting on the relatively short period of time actually devoted to discovering the structure, Crick stated:

> We had one intensive period of model building toward the end of 1951, but after that I myself was forbidden . . . to do anything further, as I was still a graduate student. For a week or so in the summer of 1952 I had experimented to see if I could find evidence for bases pairing in solution . . . The final attack [in February and March 1953] . . . only took a few weeks. Hardly more than a month or so after that our papers appeared in *Nature* . . .

The first of a series of four articles appeared in *Nature* on April 25, 1953, and described the structure of DNA and explained that it *controls the manufacture of proteins*, as suspected by Crick in 1948. The essential aspects of the double helix, as described in the articles, are shown below—the two helical chains, running antiparallel, with the sugar-phosphate backbone on the outside and the bases (adenine, thymine, guanine, and cytosine) on the inside. Because of the angles at which the DNA's chemicals bind to each other, all

Figure 18-2

DOUBLE HELIX

Backbone and Bases
(See Figure 18-1)

Backbone and Bases
in Double Helix Form

DNA molecules consist of two parallel twisting strands, like the handrails of a spiral staircase—thus, the label that immediately became famous with the Crick-Watson discovery: *The Double Helix.*

Figure 18-2 is simply the acid shown in Figure 18-1, *plus its mirror image.* Note that adenine will only "fit" into thymine, and guanine and cytosine similarly match up because of their particular chemical attractions, that is, the hydrogen bonds. This is the clearest demonstration of the need for biology and physics to join forces as biophysics in order to understand the structure of DNA. The paired bases are "stacked," as shown in Figure 18-2. Billions of stacked bases configure themselves into disclike forms that make up each chromosome of every cell of every living creature. In the interphase stage (discussed in Chapter 16) when cells are not dividing, the DNA material (chromatin) floats around in the nucleus as a double helix, but without being organized into chromosomes. When cells receive a signal in interphase and begin to divide, that DNA condenses and coalesces into the structures that we recognize in the prophase as chromosomes (as shown in Chapter 16, in Figure 16-3).

Understanding DNA

DNA Makes RNA Makes Protein

As with the other great scientific discoveries, the structure and function of DNA can be clearly understood on a conceptual basis without the detailed knowledge of the molecule's physics, chemistry, or angles of diffraction that was so important to Crick and Watson. The points about the DNA molecule being a *double* helix and that it *duplicates* itself are always emphasized in discussions about DNA, while its primary role is often not adequately explained or recognized. To understand DNA, one has to understand the capability or function that is being duplicated. Exactly what does DNA do? *It controls the design, structure, and function of the entire organism by directing the manufacture of proteins.* It can be precisely duplicated so that each new cell that's formed is exactly like the original. This function of manufacturing proteins and the ability to duplicate itself are equally applicable to a bacterium, a protozoan, or a cell from a human lung or liver.

Proteins are made solely of amino acids (as distinguished from the nucleic acids that make up DNA and RNA). Amino acids are built around the four bonds of the carbon atom. That is, carbon has a valence of four, meaning it has four unpaired electrons in its outer shell, enabling it to form such bonds and making it the most important atom and chemical element in biology. Though there are only twenty varieties of amino acids, long repetitions of multiple sequences allow for tens of thousands of combinations of amino acids into a wide variety of proteins. In fact, there are about 50,000 different types of proteins in each of our bodies, and most of those types are unique to the human species. The same twenty amino acids in 50,000 different combinations are linked together in long chains folded up on themselves.

When amino acids join together to form proteins, they serve as the basic building blocks of life. Proteins are not simply a beneficial substance we obtain from meat and other foods. They are complex molecules that exhibit an extraordinary array of properties and functions as components of structural elements such as collagen (which holds our organs together), hormones, carriers of oxygen, and antibodies (which circulate in the blood to protect us from infections), as well as being essential enzymes (which trigger chemical reactions and changes throughout the body) and catalysts in the DNA molecule itself. In view of the complexity and pervasiveness of proteins in every cell, including within the chromosomes themselves, it's not surprising that Phoebus Levene and other biochemists were convinced for many years that genes were made of proteins rather than DNA.

Figure 18-3 shows that genes are composed of DNA that forms into chromosomes in the cell's nucleus. Red blood cells have no chromosomes, and each human egg and sperm cell has 23 chromosomes. Apart from these exceptions, the nucleus of every human cell contains 46 chromosomes—that is, 22 pairs of almost identical chromosomes, plus the XY chromosomes in males and the XX chromosomes in females. Specific genes located on the Y chromosome are necessary for "maleness." Each human chromo-

Figure 18-3

DRAWING OF ELECTRON MICROGRAPH OF HUMAN CHROMOSOME NUMBER 12

DNA double helix
A-A' is identical
to B-B'

Centromere

A

Short Arms

B

"Regions" of genes

A'

Long Arms   B'

Chromosome at the end of Metaphase (see Figure 16-3) when the DNA in the cell nucleus is duplicated, condensed, and joined at the centromere, ready to separate and pass into two new cells along with the other 22 pairs of chromosomes in the nucleus

some consists of a double helix with an average of about 100 million nucleotides (that is, various sequences of the bases A, C, G, and T) in each strand. We can now more thoroughly answer the earlier question, What is a gene?: *It is a region of DNA that controls a specific hereditary characteristic, such as hair color, height, nose shape, and thousands of other traits. The particular sequence of the bases (A,C,G,T) that make up the gene usually correspond to a single protein or complementary RNA. "Gene" means an entire functional unit, including both the DNA sequences that code for the protein and the DNA sequences that "regulate" the process but do not actually code for the protein.* Each chromosome is a chain or strand of thousands of genes linked end to end. The human genome (each person's total genetic makeup) is comprised of about 100,000 genes, that is, 100,000 specific series of three-letter sequences. Specific genes are located in specific "regions" on each of the 46 human chromosomes.

Each DNA strand is about 600,000 times longer than it is wide. If straightened out it would be about *one yard long*. When the cell, nucleus, and chromosome divide, each strand then serves as a template for the formation of a new matching strand in each of the new cells because of the structure and base pairing discovered by Crick and Watson. This explains the second major aspect of DNA, the one we usually associate with the double helix: its ability to replicate itself. In other words, as the DNA is duplicated within each cell that is undergoing cell division (which is occurring by the thousands every second in our bodies), its ability to control the cell's and body's functions by directing the manufacture of proteins is also duplicated.

This brings us back to the main function of DNA: making proteins. Because the precious genes evolved so that they are protected in the cell's nucleus (as shown in Figure 16-1), working copies of the genes must be made and must be able to leave the nucleus and direct the manufacture of proteins in other parts of the cell. Thus, a type of gene "blueprint" is needed. That blueprint is made by the other nucleic acid, RNA, which is made of A, C, G, and uracil instead of thymine. RNA-polymerase is the specific enzyme that has the ability to split DNA down the middle of the "steps." In other words, it "unzips" the bases down the middle—at their hydrogen bonds—and turns the double helix into two single helices with "half steps" exposed, breaking those bonds between the two strands that link A to T and C to G.

As will be discussed in greater detail in Chapter 19, a sequence of three of these four base pairs, called a "codon," is necessary to form a "word" that stands for a specific amino acid, as shown in Table 18-4. Since the amino acids must be joined side by side to form proteins, the sequences of these three-letter codons along the DNA strands dictate the proteins that are unique to each of us.

- One or more specific three-letter sequences of the bases (each called a codon) result in the creation of each of the twenty amino acids.
- Amino acids combine in specific order to form the 50,000 different types of proteins in the human body. Each such combination of codons is a gene.
- All 100,000 human genes are configured into the 46 human chromosomes that are located in every nucleus in every cell (except red blood cells). They become coiled into this recognizable form (as shown in Figure 18-3) during cell division.

Table 18-4

GENETIC CODES SPECIFYING AMINO ACIDS

DNA triplet	RNA triplet	amino acid	DNA triplet	RNA triplet	amino acid
AAA	UUU	phenylalanine (1)	ACA	UGU	cysteine (10)
AAG	UUC		ACG	UGC	
AAT	UUA		ACC	UGG	tryptophan (11)
AAC	UUG				
GAA	CUU	leucine (2)	ATA	UAU	tyrosine (12)
GAG	CUC		ATG	UAC	
GAT	CUA				
GAC	CUG		GCA	CGU	
			GCG	CGC	
AGA	UCU		GCT	CGA	arginine (13)
AGG	UCC		GCC	CCG	
AGT	UCA	serine (3)	TCT	AGA	
AGC	UCG		TCC	AGG	
TCA	AGU				
TCG	AGC		GTA	CAU	histidine (14)
			GTG	CAC	
GGA	CCU				
GGG	CCC	proline (4)	GTT	CAA	glutamine (15)
GGT	CCA		GTC	CAG	
GGC	CCG				
TAA	AUU		TTA	AAU	asparagine (16)
TAG	AUC	isoleucine (5)	TTG	AAC	
TAT	AUA				
			TTT	AAA	lysine (17)
TAC	AUG	methionine (6)	TTC	AAG	
TGA	ACU		CCA	GGU	
TGG	ACC	threonine (7)	CCG	GGC	glycine (18)
TGT	ACA		CCT	GGA	
TGC	ACG		CCC	GGG	
CAA	GUU		CTA	GAU	aspartic acid (19)
CAG	GUC	valine (8)	CTG	GAC	
CAT	GUA				
CAC	GUG		CTT	GAA	glutamic acid (20)
			CTC	GAG	
CGA	GCU				
CGG	GCC	alanine (9)	ATT	UAA	(termination: end
CGT	GCA		ATC	UAG	of specification)
CGC	GCG		ACT	UGA	

In forming these codes, the RNA-polymerase moves along the DNA molecule, unzipping it and allowing RNA molecules that are loose in the nucleus (and have the same base sequences) to coalesce and match up along the now-exposed A, C, G, and T points of the original strands of DNA. In fact, the RNA forms an exact transcription of the DNA (except that uracil, instead of thymine, matches up with adenine). This copy is called messenger RNA, as shown in Figure 18-5.

When the RNA-polymerase gets to the "stop sign" that's at the end of every gene, it falls off along with the newly made messenger RNA, which proceeds out of the nucleus over to one of the many

Figure 18-5

FORMATION OF MESSENGER RNA

Template

Messenger RNA

Loose RNA Nucleotides

ribosomes in the cell. The ribosome reads the RNA's message (spelled out in the three-letter words) and, according to the particular sequence of bases in the codon, assembles a series of amino acids from the stockpile floating around loose in the cell. This action creates "from scratch" a particular protein "written out" in language encoded originally by the sequence of three-letter bases in the DNA that remained in the cell's nucleus. Each such new protein reflects a small portion of the long DNA strands that contain all the three-letter codes for the thousands of different proteins.

Just as the RNA-polymerase crept along the exposed DNA's G-C and A-T base pairs to create the messenger RNA, the ribosome creeps along the messenger RNA to create a protein. Step by step, every vital protein formed in our bodies is made this way. At this very moment, thousands of ribosomes in *each* cell in your body are carrying out *millions* of reactions that are causing the amino acids listed in Table 18-4 to bind into about 2,000 new protein molecules *every second*. As each protein leaves the ribosome and emerges from the cell, it has a particular folded and twisted shape, dictated by the chemical binding of the amino acids of which it is made. This shape and chemical makeup enables the 50,000 different kinds of proteins to perform their specific functions in the body.

We've now reached the climax of understanding the essence of heredity. *Since the nucleic acids (DNA and RNA) direct the manufacture of proteins, and the protein sequence is unique for each person, it is DNA that ultimately controls all hereditary characteristics.* Each three-letter code instructs the assembly of a particular amino acid (see Table 18-4) into proteins in insects, mice and humans. However, every creature is different, and every person is unique because of the particular order or sequence of those three-letter codons. In other words, the coding sequences that cause the formation of hair on a mouse are similar, but not identical, to the sequences that form hair on a human's head. Likewise, the coding sequences that result in hair being formed on two humans' heads have greater similarity to each other than the mouse's sequences, but are not identical. This is the key to understanding hereditary material and the function of DNA, and the reason molecular biologists refer to the phrase *"DNA makes RNA makes protein"* as the "central dogma."

Crick credited Watson with the discovery that led to understanding the structure of the molecule, namely, Watson's pairing of adenine with thymine and guanine with cytosine. Watson, on the other hand, gave Crick much credit for recognizing the significance of the Chargaff Ratios, concerning the relative proportions of the bases, and for recognizing the importance of pinpointing forces in the molecules by which like attracts like. That is, Crick identified the scheme for gene *replication* during cell division. Said James Watson, "Francis had the feeling that DNA replication involved specific attractive forces between the flat surfaces of the bases."

Thus, in the first *Nature* article (about how DNA controls protein production), Crick and Watson made the ultimate scientific understatement: "It has not escaped our notice that the specific pairing we have postulated immediately suggests a possible copying mechanism for the genetic material." Five weeks after publication of that first article, Crick and Watson published a second paper in *Nature*, this time on the molecule's ability to duplicate genetic material.

Years later, Crick wrote that the ideas needed to grasp the structure of DNA

> ... are ridiculously easy, since they do not violate common sense ... I believe there is a good reason for the simplicity of the nucleic acids. They probably go back to the origin of life ... At that time mechanisms had to be fairly simple or life could not have been started. The double helix is indeed a remarkable molecule. Modern man is perhaps 50,000 years old ... but DNA and RNA have been around for at least several billion years ... Yet we are the first creatures on Earth to become aware of its existence.

The Crick/Watson discovery was the culmination of eighty years of research by scores of scientists. During their eighteen-month association, Crick and Watson made their way through thirty or forty discernible steps and missteps on the road to the ultimate solution, each derived from or dependent on an existing scientific fact or theory and each attributable to a predecessor or contemporary—people like Bragg, Chargaff, Pauling, Donahue, Wilkins, and Franklin.

James Watson's brash prediction at age 22 that he would discover the secret of life, Francis Crick's desire to understand "the borderline between the living and nonliving," and his mother's prediction that "there will be plenty left for you to find out" all were fulfilled in 1953. In 1962, Crick and Watson, along with Maurice Wilkins, shared the Nobel Prize for their work.

Chapter Nineteen

The Human Genome

"As we experiment with the passages of our genetic text, we may find that certain selections . . . work better than others . . . A couple in the future might want their child to have two copies of a gene associated with musical talent, but to lose one that conferred greater susceptibility to environmental cancer."

ROBERT SHAPIRO
The Human Blueprint (1991)

The work of Crick and Watson led directly to the realization that it is possible to read and interpret the genetic plan of any organism, despite the fact that the full text of humans has three *billion* characters that make up the 100,000 genes. That realization has now culminated in our ability to decipher the exact genetic makeup of human beings, which carries with it the promise of disease prevention and other improvements in the human condition.

Knowledge of the Structure Leads to Reading the Code

Another Moon Shot

As we saw in Chapter 18, a gene carries its information through a chemical "code" or blueprint provided by the sequence of nucleotides G-C and A-T that specifies the production of proteins. In 1954 (the year after the Crick/Watson discovery), George Gamow—the same person responsible for the "nuclear droplet" theory of the atom and the person who first called his expanding universe theory the "Big Bang"—made a significant contribution to genetics. He was the first person to propose the concept of a genetic code "written" in sets of three of the four bases, called "triplets" of nucleotides. Also as seen in Chapter 18, the four bases of the nucleic acids that comprise the DNA molecule form codes that detail which of the twenty amino acids that make up all proteins will be aligned in a particular order. Using two letters at a time (out of A, C, G, and T) would allow for only 16 words. But 64 "words" or combinations can be made if *three* of the four letters are used, which is more than enough to direct "construction" of the twenty amino acids that make up proteins. Beginning with the first primordial cell, this simple "construction" function and language evolved as a result of the natural selection process described in Parts Five and Six.

Table 18-4 listed the DNA code for all 20 amino acids, including the resulting RNA complement. In humans, the body makes twelve amino acids (called nonessential amino acids), while the remaining eight (essential amino acids) must be obtained through our diet. Again, "DNA makes RNA makes protein" applies here. *With the exception of some viruses, the DNA sequence of three-letter words in all living things is copied by RNA, the genetic messenger (in a process called transcription) that directs the synthesis or "manufacture" of proteins (a process called translation).* The twenty amino acids and the three-letter codes specific to each one are the same in all living organisms, but the particular proteins that are made of the amino acids differ from one species to the next. Most types of proteins contain about 100 amino acids per protein molecule. Therefore, several types of amino acids occur repeatedly in the protein. That's the reason there are about three billion characters (A, C, G, or T) that comprise the 100,000 genes. Each cell "knows" its genetic code, thereby enabling a multicelled organism to function simply by carrying out *instructions* needed to *assemble* the "project." If one can imagine fifty instructions to assemble a piece of furniture or a working motor, "assembling" a living organism (by millions of instructions) certainly is quantitatively different, but it is conceptually the same.

About once in every ten billion times that a base (A, C, G, or T) is duplicated, the replication "machinery" makes an error by skipping or adding a letter, or by coding for the wrong letter. These are called mutations, and are copied in all future cell generations, since incorrect DNA sequences are copied as accurately as correct ones. Some mutations are "silent"—that is, they do not affect the function of the protein created by the mutated gene. As mentioned in the discussion of genetics in Chapter 17, in rare cases a mutation will create a gene with an improved function, leading to a gradual change or improvement in an entire species through the process of natural selection. In still other rare circumstances, mutations lead to changes in crucial proteins, which cause harm or death to the organism.

After Crick and Watson discovered the DNA structure, there was an explosion of work in genetics labs worldwide in an effort to read the instructions. In 1956 the British biochemist Vernon Ingram (1924–) discovered that sickle-cell anemia results from a mutation in the nucleotide sequence that codes for one single amino acid (glutamic acid) in the red protein called hemoglobin, which carries oxygen in our red blood cells. That discovery demonstrated the critical link between the code sequence and the functions of the body. It is now known that where the three-letter nucleotide "word" should have been "GAG" (which is part of the DNA/RNA code for glutamic acid, as shown in Table 18-4), it is misspelled as "GTG." This mutation in one gene, resulting from the incorrect code for one amino acid in the hemoglobin protein, changes the shape of that protein's surface, which in turn limits its ability to carry oxygen. The red blood cells that contain these abnormal hemoglobin molecules take on a deformed "sickle" shape. The resulting oxygen deficiency can lead to damaged organs throughout the body, vulnerability to infection, and death. Sickle-cell anemia is the most common genetic disease among the black population, affecting millions of people throughout the world—all because of a tiny, one-letter misspelling in the genetic text.

Mutations can also occur as a result of correctly coded genes lining up at the "wrong" location on a chromosome. As we saw in Chapter 16, during interphase and in the prophase stage of cell division, long coils of DNA are wound diffusely throughout the nucleus of the cell and begin to condense into chromosomes. It is at this stage that the individual coding sequences, the genes, can first be recognized as discrete bands on the chromosomes. Every gene lines up on the same chromosome in the same order every time. However, during DNA replication and chromosome segregation some genes that are supposed to line up on one chromosome end up on another or in the wrong place on the right chromosome. As first recognized by geneticist Edward East, these chromosomal mutations, as well as mutations in the individual three-letter codes, can also lead to an advantage or disadvantage in a cell's or organism's likelihood to survive, thus playing a crucial role in the process of natural selection.

When research by the English biochemist Frederick Sanger (1918–) enabled us to begin sequencing RNA in the 1960's, it became theoretically possible to understand all of the vast amount of information in DNA, not just isolated examples. This led to a growing interest in "decoding" the molecule so we could actually know the relationship between each gene and each bodily characteristic, including gene-based diseases. In 1975 Walter Gilbert (1932–), a biochemist at Harvard, was the first to apply a particular chemical treatment to DNA to break it into fragments and to recognize the usefulness this might have in reading the script. The following year, Sanger and his assistants at Cavendish Laboratories in Cambridge, England, spurred on by the reports of Gilbert's results, independently developed the "chain termination method" for sequencing DNA. Up to 300 base sequences could be read in a single day. Their publication of the full sequence of a particular virus's DNA (called phi-X-174) in *Nature* on February 24, 1977, was the first DNA genome—that is, the first analysis of the full genetic text of any organism. Through this new method, Sanger made it theoretically possible to determine all of the "text" that governs the heredity of any living organism, including humans. Yet, because of the much greater complexity of humans (we receive three billion characters from each parent) compared to that of a virus, much work lay ahead to make this a reality. In 1981 Sanger, Gilbert, and Paul Berg, a pioneer in cloning, received the Nobel Prize for Chemistry for their independent work in sequencing DNA.

Despite the major accomplishments of Ingram, Sanger, Gilbert, and others, the effort to decipher the code of the human genome remained fragmented until 1989, thirty-six years after Crick and Watson displayed their wood-and-metal model of DNA. In that year, a coordinated national effort, similar in scope to the Manhattan Project and the space project, was begun. None other than James D. Watson addressed a gathering of scientists in Washington, D.C., at a meeting called Genome I, which marked the consensus among molecular geneticists and others that it was time to launch a comprehensive effort to read the full text of the human genome. Such a "text" would contain the complete "blueprint" and evolutionary history of the human species. With an objective that was as clear and inspiring as landing on the moon, the Human Genome Project was born.

The Human Genome Project Is Launched

Landmarks and Maps

Watson had remained active in DNA research from 1953 to 1989, including dividing his time between Harvard and the Cold Spring Harbor Laboratory, New York, from 1968 to 1977. Francis Crick, on the other hand, ended up at the Salk Institute in San Diego to focus on his new passion (determined, no doubt, by the "gossip test")—the study of the brain and the nature of consciousness. In 1977 Watson had begun working at Cold Spring Harbor full-time as its chief administrator, greatly enhancing its academic reputation over the next decade (as well as raising funds enabling it to increase its annual budget from $600,000 to $28 million) and laying the foundation for this laboratory to become the headquarters for the Human Genome Project.

After the Genome I meeting in 1989, Congress appropriated $3 billion ($200 million a year for fifteen years) for the Genome Project and named James Watson as director of the newly created National Center for Human Genome Research, under the auspices of the U.S. Department of Health and Human Services. The project began on October 1, 1990. After much initial uncertainty over funding and other issues, it is now projected to be completed ahead of schedule, before the year 2005. Thus, we are in the midst of this monumental task of reading all the genetic messages that make us uniquely human.

By 1990 various data banks around the world contained about 50 million characters relating to various DNA molecules, about 10 percent of which pertained to human DNA. The project has made great strides in understanding the data developed during the 1980's and in accumulating and organizing human DNA data. Almost weekly news accounts relate the results of the project's ongoing work, as well as human genome discoveries made in university and commercial laboratories. The following is a small sampling of those developments:

- *December 1989:* Scientists at MIT discover a gene that they believe is crucial in the development of human immune defenses, referred to as the gene "RAG-1," which stands for "recombination activating gene." The discovery sheds new light on the complexities of the immune system, which is vital to every aspect of human health and development.
- *May 1991:* Doctors at Johns Hopkins Children's Center in Baltimore pinpoint the stage at which an error occurs in a mother's unfertilized egg, causing a child to be born with Down syndrome, the most common genetic cause of mental retardation. By analyzing a specific marker, called DNA polymorphism, they determine the origin of an extra chromosome and identify the point in time when the chromosome division malfunctions.
- *August 1991:* A collaborative research effort involving scientists from Johns Hopkins School of Medicine, the Cancer Institute in Tokyo, and the University of Utah identifies the gene that initiates colon cancer. The gene is called APC (for adenomatous polyposis coli). This discovery will allow doctors to detect a colon tumor at the earliest possible stage.
- *March 1993:* Researchers announce that Huntington's disease results from unexplained "genetic stutters," expansions in the size of a particular gene on chromosome 4, which add extra strings of the amino acid glutamine to the protein that the gene normally encodes.

- *August 1993:* Researchers at Duke University Medical Center announce that people born with a variant of a gene called APOe (for apolipoprotein E) are more likely to develop Alzheimer's disease by age 70 than people who carry other versions of the same gene.
- *June 1995:* A University of Toronto team announces that a gene on chromosome 14 is responsible for as many as 80 percent of familial cases of Alzheimer's disease.
- *August 1995:* Researchers at the University of Texas Health Science Center report that the BRCA1 gene plays a major role in breast cancer.
- *December 1995:* British scientists announce the discovery of a second gene linked to breast cancer, BRCA2.
- *February 1996:* Scientists identify the gene that encodes a variety of cell surface proteins that travel to the brain and help regulate body weight, and speculate that obesity results from a mutation in that receptor gene.
- *March 1996:* Researchers at the Oregon Health Sciences University report that healthy liver cells transplanted into diseased livers produce the missing enzyme FAH (for fumarylacetoacetate hydrolase). This gives new hope for gene therapy directed at the liver, which would reduce the need for liver transplants.
- *March 1996:* Researchers at five major medical centers announce they have found a gene that heightens the risk of kidney disease and other lupus-related disorders. The defective version of this gene codes for a protein (called an Fc receptor) that is less efficient in its immune function than a normal version of the gene.
- *April 1996:* Molecular biologists at the Seattle VA Medical Center, the University of Washington, and Darwin Molecular Corporation announce they have found the human gene that causes the symptoms of aging. They predict that as we further understand how this gene works, we could have the ability to arrest the aging process and modify the gene's involvement in triggering heart disease, cancer, and osteoporosis.

Periodically, Human Genome Project researchers publish a "map" of the human genome. They have pinpointed the physical location of over 15,000 out of the total of 30,000 "landmarks" along the strands of DNA material that comprise our chromosomes. Each of the discoveries of genes listed above relates to one or more specific sites on a particular chromosome. As stated, the defective gene that causes Huntington's disease is located on chromosome 4. In November 1996, researchers at the National Institutes of Health also found the site for Parkinson's disease on chromosome 4. Atherosclerosis (coronary artery disease) is associ-

Figure 19-1

GENOME MAP OF HUMAN CHROMOSOMES 19 AND 21

ated with a gene on chromosome 19, and Lou Gehrig's disease (amythrophic lateral sclerosis) is the result of a deficiency in a protein coded by a gene on chromosome 21. Figure 19-1 is a genome "map" of all the "regions" of genes on human chromosomes 19 and 21, with the short arm on top and the long arm on the bottom (see Figure 18-3). The numbers represent a universal system for scientists to locate such regions on strands of DNA when they have contracted into chromosomes. The black bars to the left of each chromosome depict "bands" from a special stain used to indicate the general areas on a chromosome where specific genes are located. Some of such bands contain thousands of genes.

The "Human Transcript Map" is published by the National Institutes of Health. Indeed, on NIH's home page on the Internet, one can "browse by chromosome" to see the totality of scientists' current efforts to map the human genome. As mentioned in Chapter 18, the 46 human chromosomes *average* about 100 million nucleotides each. But they are numbered according to their size—chromosome number 1 contains about 280 million nucleotides, while chromosome 22 has about 52 million. The sex chromosomes (X and Y) were not included in the numbering process. Chromosome X is about average size, and chromosome Y is one of the smallest.

The Project Spawns Hope, Fear, and Controversy

A Tale of Two Nuclei

Diseases caused by genetic mutations are a major cause of disability, death, and human tragedy throughout the world. Two percent of all newborn babies suffer from such genetic defects. Forty to fifty percent of all miscarriages are due to chromosome abnormalities. About forty percent of all infant mortality is due to genetic disease. Thirty percent of all children and ten percent of adults are admitted to hospitals at some time in their lives because of genetic disorders. Cancers are the result of a series of genetic "errors" that cause loss of control of cell growth. The Human Genome Project was originally conceived and continues to be driven primarily by the hope of curing or reducing such diseases. The recent advances in genetic research demonstrate that this expectation was well founded and confirm the practical application of this work.

Yet the project has not been without opposition. Similar to the debate that ensued after physicists harnessed the power of the atom's nucleus, unlocking the secrets in the cell's nucleus has given rise to a controversy remarkably reminiscent of that debate. The genetic code is now understood to such an extent that redesigning the human genome and directing its instructions is feasible in the near future. Many people see great potential in applying this knowledge to cure diseases and improve the human condition, while others violently oppose such genetic engineering and gene therapy on ethical and scientific grounds. Indeed, in October 1993 Robert Stillman, a fertility specialist at George Washington University Medical Center, cloned human embryos, using methods common for breeding livestock and other animals. This was a laboratory experiment and was not done in connection with a pregnancy, but it did raise the possibility that identical twins could be "created" and born years apart. It also raised enormous ethical and legal issues.

Some people fear that the line will not be drawn at curing disease and that there will be an effort to genetically engineer "perfect humans" or better health, better physical appearances, or greater intelligence. Who would make the decisions? A wide array of activists, philosophers, scientists, and religious leaders has raised the specter of sliding past the cure of Huntington's disease and cancer into "fixing" color blindness and skin color, and perhaps even causing the extinction of the human race by tampering with the gene pool and altering evolution's natural course.

Several years ago, a group made up of such diverse members as Pat Robertson, Jerry Falwell, Nobel laureate George Wald, twenty-one Roman Catholic bishops, and several Protestant and Jewish leaders issued a statement that science and society should not attempt to engineer specific traits into the human genome. They based their opposition primarily on the position that no person, group, or institution could ever "claim the right or authority to make decisions on behalf of the rest of the species alive today or for future generations." In a further link between the atom's nucleus and the cell's, this group's resolution included the statement that "Society should oppose human genetic engineering with the same courage and conviction as we now oppose the threat of nuclear extinction."

Those supporting further pioneering in genetics point out that knowledge of heredity dates back to prehistoric times and that directing or controlling the traits of offspring has been done for centuries in the breeding of plants and animals. They point to the growing number of practical applications of genetic research and the tremendous promise of using genetics to cure human diseases. Rewriting the DNA misspelling of glutamic acid in hemoglobin would cure sickle-cell anemia. Similarly, other potentially curable gene-based diseases or abnormalities include cancer, dwarfism, heart anomalies, Tay-Sachs disease, cystic fibrosis, cerebral palsy, cleft palate, and mental retardation.

It is more than theoretically possible to rewrite the genetic message in the sperm or egg, thus raising the hope of preventing hemophilia and other diseases in the newborn of afflicted parents. In January 1996 a healthy baby boy was born to a Chinese couple in New York. The birth was remarkable because when the baby was a fetus he'd been diagnosed with *alpha thalassemia*, a fatal genetic defect that occurs predominantly in Thai and Chinese people, in which the fetus is not able to make hemoglobin, which carries oxygen in the red blood cells. This defect usually leads to a miscarriage within four months after conception, but the baby survived because he underwent transplants of cells in the bone marrow *while in the womb*.

By applying our minds and the methods of science, should we take advantage of the opportunities to enhance our ability to survive, to make life longer, richer, and healthier? Expanded applications of genetic engineering, as well as experimental embryology, cloning, and transplants, promise to be the new frontier of biological and medical developments in the 21st century. Further controversy is sure to ensue.

The Uniqueness of the Code Has Practical Applications

What Did They Have on O.J.?

In addition to the goals of the Human Genome Project, DNA analysis and alteration has a myriad of other direct positive impacts on our lives. One example was mentioned in Chapter 16, namely, the research being performed by drug companies to determine the particular DNA of the new strains of tuberculosis bacteria. In addition, because the genetic code that governs all life forms is identical in all species and because we now can read that code to stimulate the manufacture of proteins, an entire industry has emerged in the last twenty years—the biotechnology industry. The three-letter instructional sequences that make up a gene from one organism (such as a plant) can be placed in another organism (bacteria are a prime example), and be actually read and followed by the other organism. As a result, genetically altered bacteria have been developed and are now used by biotechnology companies to make human growth hormones. Understanding the central dogma ("DNA makes RNA makes protein") has made it possible to help the body make critical protein molecules where it otherwise would lack that ability.

Amgen Inc. was founded in 1980 as a pioneer in the biotechnology revolution, and now markets the two largest-selling products in the industry, Epogen and Neupogen. Epogen is a genetically engineered version of a protein called erythropoietin (EPO), which stimulates the production of red blood cells in bone marrow. The drug is used primarily by dialysis patients suffering from chronic anemia as a result of their kidneys' failure to naturally produce sufficient amounts of EPO. More than 175,000 patients in the United States alone use Epogen each year, eliminating their anemia and reducing the need for blood transfusions.

Neupogen, Amgen's other major drug, stimulates neutrophils, the body's infection-fighting white blood cells. It does so because, like Epogen, it's a genetically engineered protein identical to the naturally made proteins that stimulate neutrophils. Neupogen is used by cancer patients whose natural neutrophils have been destroyed by chemotherapy. The drug not only wards off infections, it allows patients to tolerate higher doses of chemotherapy over longer periods of time, which expands treatment options and enhances their chances of surviving cancer. Amgen has twelve more DNA/protein-related drugs in clinical trials around the world. For example, in March 1996 Amgen reported that a drug called GDNF produced significant improvements in monkeys with symptoms of Parkinson's disease.

Like Amgen, other companies continue to pin their existence and future on the long-term viability of molecular biology-based drugs. Millennium Pharmaceuticals, Inc., a biotechnology firm founded in 1993, raised $180 million in research and licensing agreements to fund its effort to find genes that contribute to the development of atherosclerosis, obesity, diabetes, and asthma. In 1996 Monsanto Company received a patent for a class of genes that helps agricultural crops fight off damaging insects. Mycogen Corporation and Ciba-Geigy Corporation market similar products, which recently gave rise to patent litigation over the right to develop the genes, since the market for this gene (from an organism called *Bacillus thuringiensis*) is projected to be worth billions of dollars in the coming years.

Applying current knowledge of the genetic code, what did the Los Angeles police have on O.J. Simpson in his 1995 trial, and why has DNA become a more precise method than fingerprints to determine a person's identity? These questions take us back to the point discussed in Chapter 18—that the G-C and A-T sequences are different for each living organism. Though there are only these four "letters" in the DNA alphabet, the particular order of the long series of sentences, paragraphs, and volumes comprising the genetic story within each of us is unique. With our newfound capability to read that story, it has become a powerful piece of evidence in identifying criminals who leave behind DNA in the form of blood, semen, hair, or skin. Thus, through its expert witnesses the prosecution in the O.J. Simpson case introduced into evidence the unique sequences of Mr. Simpson's DNA from the blood found at the murder scene, in his infamous Ford Bronco, and at other locations.

Beginning in the mid-1980's, various tests were developed by molecular biologists for use in identifying individuals, not only in criminal cases, but also in missing-person cases and paternity suits. Unlike traditional hair testing, which only determines whether hair found at a crime scene has color and size characteristics consistent with the suspect's, or traditional blood and semen tests, which can pinpoint a suspect with 90 to 95 percent accuracy, DNA typing can identify a suspect or other individual with virtual certainty. The statistics for the reliability of DNA typing are generally described in a range from 1 in 5 billion to 1 in 30 billion. Since these odds are well beyond reasonable doubt, this leaves the defendant with very narrow options to refute this evidence, such as raising questions about the handling of the samples and whether the test was properly conducted. O.J. Simpson's attorneys were apparently successful in their extensive efforts to raise doubts related to such questions. Since the mid-1980's, DNA typing has gained general acceptance in the scientific and legal communities and has been used in thousands of murder, rape, burglary, and paternity cases.

The manipulation of DNA has also led to great strides in the advancement of dentistry. Periodontal (gum) disease is one of the leading causes of tooth loss. Due to the advent of DNA probes, specific bacteria can be identified and eliminated with antibacterial therapy. The key to this process is to purify specific target DNA from several suspected bacterial pathogens. The DNA strand is fragmented into small pieces and tagged with a radioactive label. The dentist then takes a sample of plaque from the patient's mouth and sends it to the laboratory for analysis. The technicians extract the bacterial DNA from the plaque sample and bind it to a filter. Similar DNA from the same bacteria will bind together when allowed to mix. Samples of DNA from the plaque are mixed with each of the radioactive bacteria DNA probes. If the suspected species of bacteria is present in the patient's sample, probe DNA will bind to it and can be viewed by exposing the sample to photographic film. The radioactive DNA will make a mark on the paper. The more bacterial DNA that is present, the darker the mark will be. The most prevalent bacteria in the plaque sample can be identified so that the proper antibiotics can be used to eliminate it.

Thus, in a process similar to identifying criminals' DNA, dentists are able to eliminate the "guilty" bacteria strains.

Genetic researchers are even experimenting with a bacterium gene that makes a substance similar to polyester and can be inserted in a cotton plant to grow a fiber that has the texture of cotton but the warmth of a heavier fabric. Cotton already has been genetically engineered to be resistant to insects, and is the subject of further studies to develop wrinkle-free and shrink-resistant strains. Also, bacterial DNA is used to produce strains of tomatoes, potatoes, corn and squash that are grown and consumed around the world that resist insects, survive herbicides and ripen slower than the normal strains.

The discovery of the structure of DNA in 1953 was the closing of the circle that began with Darwin's subtle and powerful understatement in *The Origin of Species* ninety-four years earlier: "Much light will be thrown on the origin of man and his history." Tying the origin of living things and evolution to the discovery of DNA's structure, Francis Crick stated:

> Every organism, every cell, and all the larger biochemical molecules are the end result of a long intricate process, often stretching back several billion years . . . What is found in biology is *mechanisms*, mechanisms built with chemical components and that are often modified by other, later mechanisms added to the earlier ones . . . Nature could only build on what was already there . . .

Crick was astonished that in our modern culture so few people really understand natural selection. "They don't understand it," he surmised, "because the process is very slow, thus, we rarely have any direct experience of it operating," and there appears to be a contrast and contradiction "between the highly organized and intricate results of the process—all the living organisms we see around us—and the randomness at the heart of it." Said Crick, "Some people also dislike the idea that natural selection has no foresight . . . It is the 'environment' that provides the direction . . ."

With the publication of *The Origin of Species* in 1859, Charles Darwin took the first step toward fully understanding the basic internal mechanism that explains the origin of life and the ability of organisms to reproduce. A few years later, Gregor Mendel attributed heredity to invisible and unknown "heredity factors" in living things. Less than a century later, Crick and Watson answered the question *What is a gene?* and thereby became forever entwined with the legacies of Darwin and Mendel. Today, the Human Genome Project is unlocking the incredible story contained in our genes.

Epilogue

The Synthesis

In Focus

During the one thousand long years of the Middle Ages, Western civilization was deeply entrenched in a state of mind that was hostile to rational thought. It would not have emerged from that mentality had it not been for a handful of European scholars who cleared a path that eventually led to the Renaissance. The remarkable progression of science began with the Copernican Revolution and marched through the 19th century, culminating with a great crescendo in the 20th century, the most incredible century in the history of science.

In the brief span of the last one hundred years, our understanding of life and the universe has become rich and full. Like a photograph developing in a darkroom, the image has become clear and focused. Our thoughts about the natural world have crystallized. This knowledge is embodied in the institution of science and has swept into every aspect of our daily lives. It is the foundation for the astounding technological achievements of this century.

In this epilogue we will reflect on the most incredible century, the complex institution that we call science, and the astonishing synthesis of formerly diverse disciplines that now characterizes modern science. We will touch on achievements in science as well as failures, and on scientists' limited ability to direct or control how science and technology are applied in society. We will also briefly examine the importance of freedom of expression and the value of rational thought, without which science would not exist.

Epilogue

The Synthesis

"There is one gift above all others that makes man unique among the animals . . . his immense pleasure in exercising and pushing forward his own skill . . . Discovery is a double relation of analysis and synthesis together. As analysis, it probes for what is there . . . As a synthesis, it puts the parts together in a form by which the creative mind transcends the bare skeleton that nature provides."

JACOB BRONOWSKI
The Ascent of Man (1973)

"Seemingly unrelated events, diverse scientific discoveries, industrial trends, and religious outlook can all, in historical perspective, be observed to evolve . . . before they crystallize into a new pattern . . . It is like looking into a chemical retort which is about to produce some rare and many-sided crystal. One moment everything is in solution . . . and yet in the next instant a shape has appeared out of nowhere."

LOREN EISELEY
Darwin's Century (1961)

We entered the 20th century riding horses. We will leave it riding spaceships. We entered the century dying of typhoid and smallpox, and will leave it having conquered those diseases. At the turn of the 19th century, organ transplants were unthinkable, while by the turn of this century many will have survived because another person's heart or other vital donated organ sustains them. In 1900 the human life span was 47 years. Today it is 75. We entered this century communicating over short distances with the newly invented wireless radio. We now send signals and color pictures back and forth across billions of miles of space.

We entered the 20th century without any idea of the dimensions of the universe, believing that nothing lay beyond the Milky Way Galaxy and that the stars are motionless. We will leave the century with the knowledge that we ride a tiny sphere near a star on the far reaches of a rotating galaxy among billions of other stars and galaxies, all still reeling from a violent explosion that took place 15 billion years ago.

We entered the 20th century satisfied with Newton's physics and leave it knowing that Einstein's principles are needed to fully explain the universe. As this century began, we had just begun to explore Leucippus's atom. As it comes to a close, we've broken the atom into particles with astonishing skill—one nuclear weapon can release more destructive power than all the energy of all the weapons in all the battles in the previous history of mankind.

In 1900 we suspected Darwin was right. As we approach the next millennium, we have removed all doubt as we have unearthed Lucy and other ancestors. We have come to understand the intricate workings of the cell, as we have learned to map the DNA molecule, clone animals, and decipher genetic codes.

Reason Becomes the Soul of Science

Myths in the Hands of Demagogues

In 1560 Giambattista della Porta (1535–1615), an Italian physicist, established the world's first organization dedicated specifically to the exchange of ideas among scientists. His institution, Academia Secretorum Naturae (Academy of the Mysteries of Nature), was controversial and short-lived, but signaled the dawn of the Age of Discovery in the Renaissance. In 1620 Francis Bacon (1561–1626) wrote *Novum Organum*, which was a modern version of Aristotle's *Organum*. Bacon, an English philosopher, member of Parliament, and Chancellor of England under King James I, was the first to formalize detailed rules for ensuring truth in the pursuit of science. He is credited with establishing what is now universally accepted as the scientific method. Galileo, a contemporary of Bacon, further developed and refined it.

During the Renaissance we began to recognize and accept what scientists were finding. Aristotle's and the Church's view of the universe was abandoned. Astronomers such as Tycho Brahe and Galileo made their observations and developed hypotheses. With the advent of the scientific method, hypotheses that survived scrutiny became theories, and theories that survived efforts to disprove them became laws and principles, for they must be repeatable and have the ability to survive rigorous and inevitable challenge.

In the 17th and 18th centuries the governments of Europe began to provide extensive financial support for scientific research. After Galileo's trial, it became common in Italy, England, France, and Germany to advocate science, experimentation, and theories despite the opposition of religious leaders in those countries. Thus, science had not only developed institutions and a method by which to work, it had finally gained a safe haven—a "zone of freedom" in which to grow and prosper.

At the end of the 20th century, we've now immersed ourselves in what Newton called the vast sea of knowledge. We've felt and tasted the water. We've washed it over ourselves—and what we've learned would astound him and the other great and imaginative minds of generations past. Science and its tradition of reason, applied properly by governments in a moral world society, must remain one of humankind's foremost goals. As Jacob Bronowski said, "There is one gift above all others that makes man unique among the animals . . . his immense pleasure in exercising and pushing forward his own skill." Discoveries in science have opened our minds and justified our faith in our unique abilities.

Understanding the universe and ourselves must continue to be the goal of science. In order to achieve that goal, institutions must exist that best facilitate a free and prosperous society. We must always be cognizant of how fleeting that freedom can be. Human advancement in all respects is bound up tightly with freedom. Intellectual growth is stifled when freedom of action or expression brings the threat of sanction. When change in long-held beliefs is suggested, the response of many people is to restrict the free exchange of ideas that supports such change. Thus, lack of freedom and fear of change are closely related. Insistence on the status quo is often based on myth, not a conscious suppression of truth, yet this can be more of a threat to freedom than suppression fueled by outright lies. As John F. Kennedy said at the Yale University commencement exercises in 1962, "The great enemy of truth is very often not the lie, deliberate, contrived, and dishonest—but the myth, persistent, persuasive, and realistic." Myths continue to proliferate today, and in many respects dominate thought throughout the world.

Freedom does not depend on the law, but requires the public to understand and appreciate rational thought. Reason is the soul of science. Once it is suppressed or abandoned, totalitarianism or anarchy quickly fills the void and impacts the entire community, not only science. For example, if a society abandons rational thought and allows or mandates that Creationism be taught in public schools, where would it stop? People For the American Way, a nonpartisan constitutional liberties organization based in Washington, D.C., issues an annual survey of censorship and related challenges to public education. Its 1996 report stated that there is "a renewed fight to add Creationism to science classes" and that would-be censors "are increasingly using the state legislatures in an attempt to impose their ideological . . . agendas on the nation's classrooms." Numerous legislative initiatives that recently failed would have made teaching evolution grounds for dismissal. But support is increasing, as is the likelihood of this type of bill becoming law in the near future. A society that condones such irrational laws could readily condone and support other irrational laws and social policies, such as those regarding race, color, and religion. Intellectual, artistic, and scientific progress would cease, as it did in Germany in this century, and as is the case in some totalitarian societies that exist today.

After escaping from the oppressive Soviet Union, nuclear physicist and human rights activist Andrei Sakharov (1921–1989) wrote that "Intellectual freedom is essential to human society . . . Freedom of thought is the only guarantee against an infection of people by mass myths, which, in the hands of treacherous hypocrites and demagogues, can be transformed into bloody dictatorships." Mentalities and views of the world can't be changed by decree. They can only be formed or changed by open discourse in a society that is not afraid of discarding old ways of thinking and is interested in expanding its horizons. The 20th century has been punctuated by periods of severe antirationalism. This mentality ebbs and flows throughout history, and is remarkably alive and intense as we approach the millennium, making it as imperative now as at any other time in history to cherish and support freedom of expression and our reliance on reason and rational thought.

The Progress and Application of Science Is Controlled by Our Political and Social Institutions

Glowing Blue Powder and Fourteen Billion People

In Chapter 19, we examined the controversy surrounding the human genome and genetic engineering. Debate also continues on the use of nuclear power and the atomic bomb, and the applications of other discoveries and technology. Science and scientists are often blamed for the negative impacts of such discoveries. Yet the decisions concerning the use of such knowledge are made by society as a whole primarily through governmental institutions. This includes the debate over allocation of resources. In many nations, those who want products of technology for combat and destruction compete with those who seek funds for health care and other social needs. Some of those countries have the best jet fighters, but many of their people have no toilets, no baths, no furniture, no medicines, and no education. Despite our great advances in science, much of humanity lives in ignorance and poverty. As Einstein wrote, "Rational thinking does not suffice to solve the problems of our social life." More than four hundred children die every hour from diseases such as cholera, typhoid fever, dysentery, and hepatitis. Every year millions of people, mostly children, die of starvation. Where are the blessings of technical civilization and the progress of science where these children lived?

It is not only in war that scientific discoveries and technological advances have played a role in death and destruction. On December 3, 1984, a release of the toxic gas methyl isocyanate from a plant in Bhopal, India, owned by Union Carbide Corporation eventually caused the death of over three thousand people. On April 26, 1986, the Chernobyl nuclear reactor in the Soviet Ukraine spewed a radioactive cloud across a 10,000-square-mile area of Europe, exposing five million people, and releasing two hundred times as much radiation as the atomic bombs dropped on Hiroshima and Nagasaki in 1945. Thirty-two people died immediately, and hundreds of thousands more came down with radiation-related illnesses, with some estimates stating that as many as 150,000 eventually died from such diseases. In the shadow of the concrete and steel sarcophagus that now entombs Reactor No. 4, the once-thriving Soviet "model city" of 40,000 people, Pripyat, is empty. About 270,000 people still live in dangerously contaminated areas.

In 1987, at an abandoned radiology clinic in Goiania, Brazil, a group of curious youngsters broke open a 300-pound lead capsule containing cesium 137, a radioactive substance used in cancer treatment. The glowing blue powder inside the capsule eventually contaminated 249 people, resulting in several deaths, amputations, and related illnesses.

The ultimate and most troubling paradox imaginable concerns the great progress of medical science. While it has saved so many lives, it has indirectly resulted in a population increase that now threatens the world. Look at these startling facts:

- In 1850 the world population was 1.2 billion.
- By 2000 the world population will total 6.2 billion.
- One-half of all people who have ever been born are alive today.
- In the next *fifteen* years, the world population will increase by more than the total number of people (1.2 billion) who inhabited the planet in 1850.

As mentioned earlier, the human life span has increased from 47 years in 1900 to 75 years today. *In the equivalent of two lifetimes, the world population has increased more than fivefold!* By reducing additional major risk factors, such as blood pressure, obesity, and smoking, the average life expectancy could increase to 98 or 99 years. As the elderly population increases, more people are likely to suffer from such long-term degenerative diseases as blindness, arthritis, Alzheimer's disease, sensory impairment, and osteoporosis, placing greater strain on every nation's economy. The world population is increasing by three people every second. Every three years the world population increases by about 275 million people—the present population of the United States—with the largest increases in the poorest countries, which are the least equipped to meet the needs of the new arrivals and invest in their future. Experts warn that unless the trend is reversed soon, the world's population will reach 14 billion people in a hundred more years.

Rapid population growth damages the economic gains many countries have achieved, holding back or crippling altogether the prospects of balanced development. It's not the absolute population that a country achieves, it is the rate of change that prevents that country from maintaining agricultural productivity, creating new jobs, and delivering health care and the other advances needed. With the exploding world population comes greater potential for violence and social upheaval. Thus, birth control and family planning are critical elements in achieving world peace. As a result of complicated social and political factors, half the world's women have no access to family planning services. Yet, women are the audience that must be reached and educated if developing nations are to successfully stem population growth. H.G. Wells once said that civilization is engaged in a race between education and catastrophe. Catastrophe, in the form of runaway population growth, is winning.

However, the scientists' role in the world's social ills is only indirect. The solution is not in science. Our failings are not in science. They are found in the social and cultural structures and institutions we've created because politics, government, and our own morality determine how science and technology will be *applied*. Einstein noted that "Science can only ascertain what *is*, not what should be . . . Outside of its domain value judgements of all kinds remain necessary." Robert Oppenheimer, reflecting on his own role in whether to drop the atom bomb, pointed out that as director of the Manhattan Project, "I had no policy-making authority." Similarly, the scientists involved in the Human Genome Project are working on scientific advances, not social policy.

The object of science and technology is not to conquer nature or to conquer people. For Kepler and Newton, it was to understand how the universe works. For Watson and Crick, it was to understand how life works. Marie Curie called science the most sublime manifestation of the human spirit and a perfect tool to aid humanity. As an institution, science seeks primarily to comprehend the physical world in a logical and uniform system of thought.

Diverse Fields of Science Merge in the 20th Century

Fertility of Hybrid Species

Loren Eiseley was referring to the history of evolutionary theory in the passage quoted at the beginning of this epilogue. But his statement about diverse scientific discoveries crystallizing into a new pattern is equally applicable to all of science in the 20th century. In this century, all the branches and fields of science coalesced into one overlapping, interwoven, intertwined, and interdependent complexity of understanding that is now hurling us into the future. The borderlines that used to clearly separate astronomy, physics, and biology have become fertile fields of common interest and pursuit, together with other fields and subdisciplines. As we saw in Part Seven about DNA, 20th-century physicists turned to biology, and biologists and geneticists found their answers in chemistry. In the 20th century, astronomers became dependent on physics and chemistry, while geology revealed key aspects of the origin and evolution of life. "One moment everything was in solution," wrote Eiseley, ". . . in the next instant a shape has appeared out of nowhere."

Twentieth-century scientists are primarily experimenters, not theorists—they are molecular geneticists, biochemists, biophysicists, cosmologists, and experts in other *combinations* of expertise not fathomed in earlier centuries. In the 1500's, Copernicus made his twenty-seven observations and then applied logic to arrive at heliocentricity. Similarly, Newton, Darwin, and Einstein epitomize the lone scientist who amassed the evidence and mulled and formulated it until the theories were shaped. Four of the five great discoveries made in the 20th century—atom, Big Bang, cell, and DNA—were made possible because "one-person science" has ended and "group science" is upon us. In the 20th century, science became symbiotic and overlapping and scientists no longer desired to maintain their isolation within a defined field. They recognized the need for cross-fertilization—the need to follow developments in the other fields on which they depended and to share information.

We saw the first signs of the connection between astronomy and physics in the early 1600's when Kepler discovered his laws of planetary motion. The final synthesis of those fields occurred in this century, as the Big Bang theory took shape. In the first decades of this century, the fields of biophysics, biochemistry, and molecular genetics developed. Beginning in 1936, George Gamow developed the first theory linking small-scale nuclear processes to the Big Bang. Biologists, chemists, and geneticists have pooled their knowledge in a myriad of ways that will remain linked forever. Carbon, nitrogen, oxygen, and phosphorus are the principal building blocks of life. Protein, which is essential to all life, is dependent on the capture of nitrogen from the atmosphere. Nitrogen is also an essential component of DNA. Further examples touch many of the other elements—sodium and potassium are critical elements of the nervous system and brain, and calcium is the most essential component in the maintenance of the form of organisms in bones and shells.

In his famous book *A Brief History of Time*, physicist Stephen Hawking tied atomic physics to biology, explaining that if the electric charge of the electron had been only slightly different, either the nuclear fusion process that transforms hydrogen into helium in stars would not have been possible, or else many stars wouldn't have exploded. In either case, without sunlight or the heavier chemical elements that are made in stars and flung into space when stars explode, life as we know it on Earth could not have developed.

In reminiscing on the history of classical genetics, Francis Crick observed that "the important thing was to combine it with biochemistry." In his book *What Mad Pursuit*, in which he discusses the process of scientific discovery, Crick describes geneticists' and protein chemists' belated recognition of their need to become more familiar with each other's field to properly conduct experiments designed to solve the coding problem of the DNA molecule. As late as 1954 (after the discovery of the double helix), experts in genetics remained extremely hesitant to spend time learning about protein chemistry, and vice versa. As discussed in Chapter 19, the chemist Vernon Ingram then identified a specific alteration in the code for one amino acid in the protein hemoglobin as being responsible for sickle-cell anemia. Crick credited Ingram's research with "a drastic change of attitude" among geneticists and protein chemists, and noted the lesson to be learned about the synthesis of diverse scientific disciplines. "In nature," stated Crick, "hybrid species are usually sterile, but in science the reverse is often true. Hybrid subjects are often astonishingly fertile . . . If a scientific discipline remains too pure it usually wilts." This hybridization has resulted in most of the extraordinary medical benefits developed in the 20th century.

Twentieth-Century Science Becomes the Science of Incremental Advances

Anonymity and the Nobel

The refinement and greater understanding of scientific principles has led to much greater specialization among scientists. Each new issue, each scientific question posed, each theory, is now dependent on thousands of scientists driving toward a collective creation, yet with each of those people concentrating on his or her narrow specialty. Thus, contrary to the seven great and broad sweeping theories that have revolutionized science, the remainder of 20th-century science has become the science of *incremental advances*—that is, narrow and complicated discoveries derived from experiments and studies in research laboratories all over the world, all contributing to an ultimate understanding of some limited phase of the physical universe or of living organisms. As Louis Pasteur once said, "Science advances through tentative answers to a series of more and more subtle questions which reach deeper and deeper into the essence of natural phenomena."

The emphasis in science today is on research and experiments to refine discrete aspects of proven discoveries. Also, it is probable there are few scientific discoveries remaining that would be comparable in significance to the seven reviewed in this book. In *The End of Science*, John Horgan predicts that while "this is the age in which we are discovering the fundamental laws of nature . . . that day will never come again." Twentieth-century science has become the science of the infinitesimal, rather than the science of the infinite. Even the science of all matter in the universe—the Big Bang—has become the science of subatomic particles, and particle accelerators, as scientists search for the "Grand Unified Theory" that seeks to tie quantum theory to the cosmos. This is the theory for which Albert Einstein searched in vain for forty years, one that would combine chemistry, electromagnetism, mechanics, astrophysics, and the rest of physics into one set of equations that explains everything.

Twentieth-century science is the science of the electron microscope and the Hubble Space Telescope. It is the science of privately funded laboratories, and thousands of university and government laboratories engaged in fundamental research, not the science of a naturalist collecting specimens of finches in the Galapagos Islands. Brookhaven National Accelerator Laboratory in Upton, Long Island, is one of three national particle physics labs in the United States. It has a 5,000-acre campus where over 3,000 people conduct research in this esoteric field, centered around the giant particle accelerator. The Research Triangle that sits between Raleigh, Durham, and Chapel Hill, North Carolina, is the home of the National Institute of Environmental Health Sciences (part of the U.S. Department of Health and Human Services) and employs about 300 people on a campus that spreads over the wooded hills. Among the types of research carried out there are efforts to determine how chemicals in our environment cause cancer and miscarriages, and how expression of certain genes leads to a variety of cancers. At Cold Spring Harbor Laboratory in New York, where James Watson worked for over twenty years, the fundamental process governing gene expression is being explored at the headquarters of the Human Genome Project.

Twentieth-century science interweaves individual contributions on a worldwide scale into the science of the multitudes, rather than the science of the one. Breakthroughs and discoveries are made by groups, teams, laboratories, and universities, rather than individuals. Even the naming of new elements added to the periodic table, which used to be the privilege of the discoverer, is being usurped by a committee of the International Union for Pure and Applied Chemistry.

Despite the transition to group science, individual achievements are recognized and rewarded in many ways, particularly through the Nobel Prize, the crowning achievement in fundamental science. The Nobel Prize, first awarded in 1901, is paid from a fund established in 1896 by the will of the Swedish chemist and engineer Alfred Bernhard Nobel (1833–1896), who invented dynamite.

Figure E-1

The Six Levels of Scientific Achievement

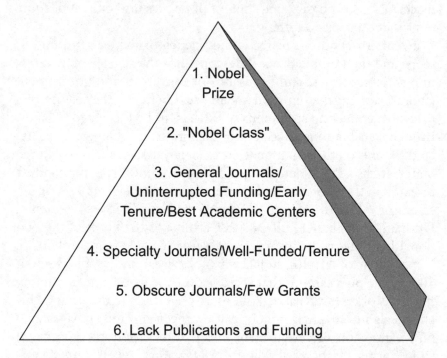

1. Nobel Prize

2. "Nobel Class"

3. General Journals/ Uninterrupted Funding/Early Tenure/Best Academic Centers

4. Specialty Journals/Well-Funded/Tenure

5. Obscure Journals/Few Grants

6. Lack Publications and Funding

Nobel's will specified that the awards should be made annually "to those who, during the preceding year, shall have conferred the greatest benefit on mankind" in three fields of science: physics, chemistry, and physiology or medicine. Nobel Prizes are also awarded annually in literature, peace, and economics.

The pyramid in Figure E-1 depicts one view of the multiple levels of scientific achievement and the scientific community. At the pyramid's apex are the winners of the Nobel Prize, people such as Curie, Einstein, Crick, and Watson. In the second tier are the "Nobel Class"—those whose work is just as imaginative, groundbreaking, and important to humanity as the Nobel winners, but who weren't the first to make a particular scientific breakthrough. Level three is occupied by those who publish solely in the best general science journals, such as *Science*, *Nature*, and *Cell*. They enjoy uninterrupted funding from federal government agencies such as the National Institutes of Health and the National Science Foundation, and from private foundations. Level four scientists are well known in their specialties and have made significant advances, publishing routinely in the best journals of their fields of study and occasionally in the most prestigious general journals. Members of the fifth level publish, but usually in obscure and sparsely read journals. They rarely make a significant impact in their fields and have difficulty attracting grant money to support their pursuits. Finally, the individuals on the sixth rung of scientific pursuit generally do not produce noteworthy papers or receive grants which would enable them to perform research important to science. Those in the pyramid number in the tens of thousands around the world, and a large number of them are making the incremental advances in science.

The existence of group science, while still recognizing the value of individual achievement, has resulted in a paradox. Despite the difficulty of their achievements, even the greatest 20th-century scientists—the ones on the top levels of the pyramid—are anonymous and largely unknown to the general public. As we have seen in the preceding chapters, Marie Curie, Ernest Rutherford, Albert Einstein, Niels Bohr, Edwin Hubble, Thomas Morgan, Francis Crick, James Watson, Arno Penzias, and Robert Wilson were all 20th-century scientists and giants in their fields. Indeed, Crick, Watson, Penzias, and Wilson are still active in their fields. But in the average household, the people on the long list of Nobel Prize recipients since 1901 are hardly known, except for Einstein and perhaps Curie.

The last of the seven great scientific discoveries was made in 1953 by Crick and Watson. *Are there no great scientists in the latter half of the 20th century?* Yes, there are great scientists, but the opportunities for mass recognition have diminished because their work, measured in incremental advances, is so complex. Though Nobel Prizes are given to individuals each year, the incremental advance for which each of those people receives the prize will continue to be consumed by the intricate labyrinth of 20th- and 21st-century science.

The Seven Greatest Scientific Discoveries in History . . .

While we continue to argue over political structures and religious philosophies, and while we are not sure how best to manufacture and distribute the world's goods or allocate wealth among people, we now know a great deal about what exists. We find ourselves with more revealed than could have been conceived in our wildest dreams of just a few centuries past.

All the matter in the universe still reels from an explosion that occurred 15 billion years ago. During the first ten and a half billion years of the universe's existence, this planet coalesced out of dust and gases, then gave birth to the first cell. It took over three billion years for that cell to evolve into the first multicelled organisms. Plants evolved that swallow photons from the Sun and mix them with water and carbon dioxide into a carbohydrate brew, while the animals inhale the plants' waste—oxygen—and return their own discarded molecules, the carbon dioxide so vital to the plants' survival. All the living organisms on this journey are the natural products of this rotating and changing sphere that revolves around an ordinary star 93 million miles away in an outer spiral arm of the Milky Way Galaxy, amid billions of other galaxies.

Another few hundred million years elapsed between the appearance of the first mammal and *Australopithecus*. During the next three million years, *Australopithecus* changed from a figure in the landscape into *Homo sapiens*, shaper of the landscape, a being intelligent enough to wonder at the universe and understand that we are six billion individuals made up of faint, elusive and finely adjusted electrical charges, configured into 60 trillion cells and pulled down firmly to this Earth by the interplay of those charges.

. . . And the People Who Made Them

After standing on the shoulders of Kepler, Galileo, and others, Newton led humankind to the edge of the vast sea of knowledge. Ernest Rutherford, Marie Curie, and Niels Bohr discovered Leucippus's ultimate particle. Albert Einstein identified the enormous power hidden within that particle, and redefined time and space.

Edwin Hubble peered through the telescope and saw the galaxies blowing apart. Charles Darwin observed the vast diversity of plant and animal life and saw a common ancestor. Through the microscope, Schleiden, Schwann, Flemming, and Weismann came to the realization that they were viewing the source of all life and growth. In generations of peas, Gregor Mendel saw undetectable heredity factors being passed on through some unknown process. East, Morgan, and others identified that process. Francis Crick and James Watson opened the secret text of life, the first words of which were written in the first cell, four billion years ago.

Knowledge is our destiny. *Homo sapiens* will continue to search for the answers to new questions. We will develop new concepts, new theories, and we will continue our quest to understand the natural world. We are different from the other animals. We must continue to discover, create, explore, and invent. We must search for the cure and the life-saving solution, for we are the discoverers, creators, explorers, and inventors. We seek the unknown—the deep, the dark, the never before seen—and we have within us the capacity for even greater wisdom.

We have come to the future. We have found our place by looking back and understanding history. We are poised to become twenty-first centurions. As Orville Wright stated, "We don't have to look too far to see the future. We can already see it will be magnificent."

We have now traveled the 15-billion-year journey.

CHRONOLOGY

1800

<Prior to 1800>

1. GRAVITY/PHYSICS	Copernicus' The Revolutions (1543) Galileo's Dialogue on the Two Chief World Systems (1632) Kepler's Laws of Planetary Motion (1609–18) Newton's Principia (1687)	
2. ATOM	Boyle's Skeptical Chymist (1661) Lavoisier identifies oxygen (1783)	Dalton's atomic theory (1803) Mendeleev's Periodic Table (1869)
3. RELATIVITY		Maxwell's four laws/electromagnetic field (1861)
4. BIG BANG	Herschel discovers Milky Way (1785)	Maxwell's four laws/electromagnetic field (1861)
5. EVOLUTION	Linnaeus' Systema (1735) Hutton founds geology (1785)	Darwin's Origin of Species (1859) Lyell's Principles of Geology (1830)
6. CELL/GENETICS	Leeuwenhoek views blood cells (1673)	Brown observes cell nucleus (1831) Virchow's Cellular Pathology (1858) Schleiden and Schwann develop cell theory (1838–39) Mendel's principles of heredity (1865)
7. DNA		

1900   2000

Rontgen discovers   Einstein's equation   Chadwick discovers   Meitner describes
X-rays (1895) $E = mc^2$ (1907)   neutron (1932)   chain reaction (1939)

Thomson discovers   Rutherford describes   Cockcroft/Walton   Fermi achieves
Electrons (1897)   atom (1912)   achieve fission (1932) large scale chain
                                                                  reaction (1942)

Curie discovers   Bohr develops   Einstein's letter   Atomic bombs on
Radioactivity (1898) quantum theory (1913)   to FDR (1939)   Hiroshima and Nagasaki (1945)

Einstein's Special Theory
of Relativity (1905)

Einstein's General Theory
of Relativity (1915)

Hubble determines   Penzias and Wilson discover
Andromeda is galaxy (1923)   cosmic background
                                              radiation (1965)

Hubble discovers recession   Discovery of possible
of galaxies (1927)   life on Mars (1996)

                                              Johanson discovers
                                              "Lucy" (1974)

Darwin's Descent   Discovery of sea-floor
of Man (1871)   spreading and plate
                                              tectonics (1963)

Discovery of hominid
fossils (1920–present)

Flemming discovers   Mendel's principles
chromosomes (1879)   rediscovered (1900)

Weismann describes   Morgan discovers recombination and   Discovery of possible
meiosis (1886)   identifies genes (1915)   life on Mars (1996)

Fisher/Haldane/Wright
develop population genetics (1920)

Levene identifies   Wilkins/Franklin X-ray   Sanger sequences
nucleic acids (1909)   diffraction (1945–1953)   RNA (1960)

Avery proves DNA   Crick/Watson discover
contains genes (1944)   DNA structure (1953)

Chargaff determines   Human Genome
A-T/C-G ratios (1950)   Project begins (1989)

Bibliography

PART ONE

GRAVITY AND THE BASIC LAWS OF PHYSICS

Andrade, E.N. da C. *Sir Isaac Newton* (New York: Doubleday, 1958).

Augros, Robert M., and George N. Stansciu. *The New Story of Science* (New York: Bantam Books, 1984).

Baumgardt, Carola. *Johannes Kepler, Life and Letters* (New York: Philosophical Library, 1951).

Caspar, Max. *Kepler* (New York: Abelard-Schuman, Ltd., 1959).

Christianson, Gale E. *In the Presence of the Creator* (New York: The Free Press, 1984).

Clarke, Arthur C. *The Promise of Space* (New York: Berkley Books, 1985).

Cohen, I. Bernard. *Introduction to Newton's Principia* (Cambridge, MA: Harvard University Press, 1978).

Cooper, Lane. *Aristotle, Galileo, and The Leaning Tower of Pisa* (New York: Ithaca Press, 1935).

Drake, Stillman. *Discoveries and Opinions of Galileo* (New York: Doubleday Anchor Books, 1957).

Drake, Stillman. *Galileo: Pioneer Scientist* (Toronto: University of Toronto Press, 1990).

Drake, Stillman. *Galileo Studies* (Ann Arbor, MI: The University of Michigan Press, 1970).

Drake, Stillman. *Galileo at Work* (Chicago: The University of Chicago Press, 1978).

Ferris, Timothy, ed. *The World Treasury of Physics, Astronomy, and Mathematics* (Boston: Little, Brown and Company, 1991).

Koestler, Arthur. *The Watershed* (Lanham, MD: University Press of America, 1960).

Koyre, Alexandre. *Newtonian Studies* (Cambridge, MA: Harvard University Press, 1965).

Motz, Lloyd, and Jefferson Hane Weaver. *The Concept of Science* (New York: Plenum Press, 1988).

Sagan, Carl. *Cosmos* (New York: Random House, 1980).

Sagan, Carl. *Murmurs of Earth* (New York: Ballantine Books, 1978).

Thoren, Victor E. *The Lord of Uraniborg* (New York: Cambridge University Press, 1990).

Westfall, Richard S. *The Life of Isaac Newton* (New York: Cambridge University Press, 1993).

Ziegler, Philip. *The Black Death* (Wolfeboro Falls, NH: Alan Sutton Publishing Inc., 1991).

PART TWO

THE STRUCTURE OF THE ATOM

Asimov, Isaac. *Understanding Physics: The Electron, Proton, and Neutron* (New York: New American Library, 1966).

Atkins, P.W. *The Periodic Kingdom* (New York: Basic Books, 1995).

Brady, James E., and John R. Holum. *Fundamentals of Chemistry* (New York: John Wiley & Sons, 1984).

Crease, Robert P., and Charles C. Mann. *The Second Creation* (New York: Macmillan Publishing Company, 1986).

Curie, Eve. *Madame Curie* (New York: Doubleday & Co., 1937).

Davies, Paul. *God and the New Physics* (New York: Simon & Schuster, 1983).

Feinberg, J.G. *The Story of Atomic Theory and Atomic Energy* (New York: Dover Publications, 1960).

Feynman, Richard P. *"Surely You're Joking, Mr. Feynman!"* (New York: Bantam Books, 1986).

Gamow, George. *Atomic Energy in Cosmic and Human Life* (New York: Macmillan, 1947).

Holden, Alan. *The Nature of Atoms* (New York: Oxford University Press, 1971).

Ihde, Aaron J. *The Development of Modern Chemistry* (New York: Dover Publications, 1984).

Malone, Leo J. *Basic Concepts of Chemistry* (New York: John Wiley & Sons, 1981).

Morris, Richard. *The Edges of Science* (New York: Prentice Hall Press, 1990).

Nitske, W. Rob. *Wilhelm Conrad Rontgen* (Tucson: The University of Arizona Press, 1971).

Packard, Edward. *Imagining the Universe* (New York: Perigee Books, 1994).

Pflaum, Rosalynd. *Grand Obsession: Madame Curie and Her World* (New York: Doubleday, 1989).

PART THREE

THE PRINCIPLE OF RELATIVITY

Bondi, Hermann. *Relativity and Common Sense* (New York: Dover Publications, 1964).

Clark, Ronald W. *Einstein, The Life and Times* (London: Hodder and Stoughton, 1973).

Einstein, Albert. *Ideas and Opinions* (New York: Crown Publishers, 1954).

Einstein, Albert. *The World As I See It* (New York: Philosophical Library, 1956).

Einstein, Albert, and Leopold Infeld. *The Evolution of Physics* (New York: Simon and Schuster, 1938).

Frank, Philipp. *Einstein, His Life and Times* (New York: Alfred A. Knopf, 1947).

Hawking, Stephen. *Black Holes and Baby Universes* (New York: Bantam Books, 1993).

Hawking, Stephen W. *A Brief History of Time* (New York: Bantam Books, 1988).

Jeans, James. *Physics and Philosophy* (New York: Dover Publications, 1981).

Kern, Stephen. *The Culture of Time and Space 1880–1918* (Cambridge, MA: Harvard University Press, 1983).

Michelmore, Peter. *Einstein, Profile of the Man* (New York: Dodd, Mead & Co., 1962).

Pais, Abraham. *'Subtle Is the Lord . . .', The Science and the Life of Albert Einstein* (New York: Oxford University Press, 1982).

Russell, Bertrand. *The ABC of Relativity* (New York: New American Library, 1985).

Sartori, Leo. *Understanding Relativity* (Berkeley: University of California Press, 1996).

Sayen, Jamie. *Einstein in America* (New York: Crown Publishers, 1985).

Schilpp, Paul Arthur, ed. *Albert Einstein: Philosopher-Scientist* (New York: The Library of Living Philosophers, 1949).

Swenson, Lloyd S., Jr. *Genesis of Relativity* (New York: Burt Franklin & Co., 1979).

PART FOUR

THE BIG BANG AND THE FORMATION OF THE UNIVERSE

Asimov, Isaac. *The Universe* (New York: Avon Books, 1966).

Barrow, John D. *The Origin of the Universe* (New York: Basic Books, 1994).

Davies, Paul. *The Last Three Minutes* (New York: Basic Books, 1994).

Fall, S.M., and D. Lynden-Bell, eds. *The Structure and Evolution of Normal Galaxies* (Cambridge, England: Cambridge University Press, 1981).

Hazen, Robert M., and James Trefil. *Science Matters* (New York: Doubleday, 1991).

Heiserman, David. *Radio Astronomy for the Amateur* (Blue Ridge Summit, PA: TAB Books, 1975).

Heuer, Kenneth. *The End of the World* (New York: Rinehart & Company, 1953).

Hey, J.S. *The Evolution of Radio Astronomy* (New York: Science History Publications, 1973).

Hodge, Paul W. *Atlas of the Andromeda Galaxy* (Seattle: University of Washington Press, 1981).

Hodge, Paul W. *Galaxies* (Cambridge, MA: Harvard University Press, 1986).

Islam, Jamal N. *The Ultimate Fate of the Universe* (Cambridge, England: Cambridge University Press, 1983).

Kaufmann, Williams J., III. *Galaxies and Quasars* (San Francisco: W.H. Freeman and Company, 1979).

Kuhn, Ludwig. *The Milky Way* (New York: John Wiley & Sons, 1982).

Longair, Malcolm S. *The Origins of the Universe* (Cambridge, England: Cambridge University Press, 1991).

Macvey, John W. *Where Will We Go When the Sun Dies?* (New York: Stein and Day, 1983).

Morris, Richard. *The Fate of the Universe* (New York: PEI Books, 1982).

Motz, Lloyd. *The Universe* (New York: Charles Scribner's Sons, 1975).

Parker, Barry. *Colliding Galaxies* (New York: Plenum Press, 1990).

Parker, Barry. *Invisible Matter and the Fate of the Universe* (New York: Plenum Press, 1989).

Piddington, J.H. *Radio Astronomy* (New York: Harper & Brothers, 1961).

Ronan, Colin A. *The Natural History of the Universe* (New York: Macmillan, 1991).

Silk, Joseph. *The Big Bang* (San Francisco: W.H. Freeman and Co., 1980).

Sullivan, W.T., III., ed. *The Early Years of Radio Astronomy* (New York: Cambridge University Press, 1984).

Sutton, Christine, ed. *Building the Universe* (New York: Basil Blackwell Ltd., 1985).

Verschuur, Gerrit L. *The Invisible Universe* (London: The English University Press Ltd., 1974).

Verschuur, Gerrit L. *The Universe Revealed* (New York: Springer-Verlag, 1987).

Weinberg, Steven. *The First Three Minutes* (New York: Basic Books, 1977).

PART FIVE

EVOLUTION AND THE PRINCIPLE OF NATURAL SELECTION

Angela, Piero, and Alberto Angela. *The Extraordinary Story of Human Origins* (New York: Prometheus Books, 1993).

Burkhardt, Frederick, and Sydney Smith, ed. *The Correspondence of Charles Darwin*, 9 Vols. (New York: Cambridge University Press, 1985).

Bronowski, Jacob. *The Ascent of Man* (Boston: Little, Brown and Company, 1973).

Campbell, Bernard G. *Humankind Emerging* (Boston: Little, Brown and Company, 1985).

Cox, Allan, ed. *Plate Tectonics and Geomagnetic Reversals* (San Francisco: W.H. Freeman and Co., 1973).

Darwin, Charles. *The Descent of Man* (Princeton, NJ: Princeton University Press, 1981).

Darwin, Charles. *The Origin of Species* (New York: Macmillan Publishing Co., 1962).

Dennett, Daniel C. *Darwin's Dangerous Idea* (New York: Simon & Schuster, 1995).

Dockens, Richard. *River out of Eden* (New York: Basic Books, 1995).

Durant, John, ed. *Darwinism and Divinity* (New York: Basil Blackwell Ltd., 1985).

Eiseley, Loren. *Darwin's Century* (New York: Doubleday & Company, 1958).

Eldredge, Niles. *The Monkey Business* (New York: Pocket Books, 1982).

Eldredge, Niles, and Ian Tattersall. *The Myths of Human Evolution* (New York: Columbia University Press, 1982).

Erickson, Jon. *Plate Tectonics* (New York: Facts on File, 1992).

Fagan, Brian M. *The Journey From Eden* (London: Thames and Hudson Ltd., 1990).

Gould, Stephen Jay. *Wonderful Life* (New York: W.W. Norton & Company, 1989).

Humphrey, Nicholas. *A History of the Mind* (New York: Simon & Schuster, 1992).

Johanson, Donald, and James Shreeve. *Lucy's Child* (New York: William Morrow and Company, 1989).

Kearey, Philip, and Frederick J. Vine. *Global Tectonics* (London: Blackwell Scientific Publications, 1990).

Leakey, Richard. *The Origin of Humankind* (New York: Basic Books, 1994).

McCrone, John. *The Ape That Spoke* (New York: William Morrow and Company, 1991).

Morris, Desmond. *The Illustrated Naked Ape* (New York: Crown Publishers, 1967).

Poirier, Frank E. *Understanding Human Evolution* (Englewood Cliffs, NJ: Prentice Hall, 1990).

Reeder, John. *Man on Earth* (Austin: University of Texas Press, 1988).

Sagan, Carl. *Broca's Brain* (New York: Ballantine Books, 1979).

Sagan, Carl. *The Dragons of Eden* (New York: Ballantine Books, 1977).

Sagan, Carl, and Ann Druyan. *Shadows of Forgotten Ancestors* (New York: Random House, 1992).

Weiner, Jonathan. *Planet Earth* (New York: Bantam Books, 1986).

PART SIX

THE CELL AND GENETICS

Leadbetter, Edward R., and J.S. Poindexter. *Bacteria in Nature* (New York: Plenum Press, 1985).

Moody, Paul A. *Genetics of Man* (New York: W.W. Norton & Company, 1967).

Morgan, Thomas H. *Heredity and Sex* (New York: Columbia University Press, 1913).

Peacocke, Arthur. *God and the New Biology* (San Francisco: Harper & Row, 1986).

Rensberger, Boyce. *Biology* (New York: Fawcett Columbine, 1996).

Rosen, Robert. *Life Itself* (New York: Columbia University Press, 1991).

Wright, Sewall. *Evolution and the Genetics of Populations*. Vol. 1 (Chicago: The University of Chicago Press, 1968).

PART SEVEN

THE STRUCTURE OF THE DNA MOLECULE

Alberts, Bruce, Dennis Brahe, Julian Lewis, Martin Raff, Keith Roberts, and James D. Watson. *Molecular Biology of the Cell* (New York: Garland Publishing, 1994).

Crick, Francis. *What Mad Pursuit* (New York: Basic Books, 1988).

Gribbin, John. *In Search of the Double Helix* (New York: McGraw-Hill Book Company, 1985).

Kornberg, Arthur. *DNA Replication* (San Francisco: W.H. Freeman and Company, 1980).

Shapiro, Robert. *The Human Blueprint* (New York: St. Martin's Press, 1991).

Watson, James D. *The Double Helix* (New York: Penguin Books, 1968).

ADDITIONAL SOURCES

Adler, Mortimer J., ed. *Encyclopaedia Britannica,* 29 Vols. (Chicago: Encyclopaedia Brittanica, Inc., 1996).

Bernal, J.D. *Science in History.* 4 Vols. (Cambridge, MA: The MIT Press, 1969).

Boorstin, Daniel J. *The Discoverers* (New York: Random House, 1983).

Brennan, Richard P. *Dictionary of Scientific Literacy* (New York: John Wiley & Sons, 1992).

Bronowski, Jacob. *A Sense of the Future* (Cambridge, MA: The MIT Press, 1977).

Burke, James. *Connections* (Boston: Little, Brown and Company, 1978).

Burke, James. *The Day the Universe Changed* (Boston: Little, Brown and Company, 1985).

Campbell, Joseph. *The Power of Myth* (New York: Doubleday, 1988).

Capra, Fritjof. *The Turning Point* (New York: Bantam Books, 1982).

Carnegie Library of Pittsburgh. *The Handy Science Answer Book* (Detroit: Visible Ink Press, 1994).

Clarke, Arthur C. *Profiles of the Future* (New York: Holt, Rinehart and Winston, 1984).

Cohen, I. Bernard. *Revolution in Science* (Cambridge, MA: Harvard University Press, 1985).

Eiseley, Loren. *The Man Who Saw Through Time* (New York: Charles Scribner's Sons, 1973).

Feldman, Anthony, and Peter Ford. *Scientists & Inventors* (New York: Facts on File, 1979).

Gardner, Martin, ed. *Great Essays in Science* (Buffalo, NY: Prometheus Books, 1994).

Gillispie, Charles Coulston. *The Edge of Objectivity* (Princeton, NJ: Princeton University Press, 1960).

Goldstein, Thomas. *Dawn of Modern Science* (Boston: Houghton Mifflin Company, 1980).

Golob, Richard, and Eric Brus, eds. *The Almanac of Science and Technology* (New York: Harcourt Brace Jovanovich, Inc., 1990).

Hagihara, Nobutoshi, ed. *Experiencing the Twentieth Century* (Tokyo: University of Tokyo Press, 1985).

Hart, Michael H. *The 100* (New York: A&W Visual Library, 1978).

Hutchins, Robert Maynard, ed. *Great Books of the Western World,* 60 Vols. (Chicago: Encyclopaedia Britannica, 1986).

Ingpen, Robert, and Philip Wilkinson. *Encyclopedia of Events That Changed the World* (New York: Penguin Books, 1991).

Knight, David. *The Age of Science* (New York: Basil Blackwell Ltd., 1986).

Kohlstedt, Sally Gregory, and Margaret W. Rossiter, eds. *Historical Writing on American Science* (Baltimore: Johns Hopkins University Press, 1985).

Kuhn, Thomas S. *The Structure of Scientific Revolutions* (Chicago: The University of Chicago Press, 1970).

Medawar, Peter. *The Threat and the Glory* (New York: HarperCollins, 1990).

Moyers, Bill. *A World of Ideas* (New York: Doubleday, 1989).

Patterson, Orlando. *Freedom* (New York: HarperCollins Publishers, 1991).

Porter, Roy, ed. *Man Masters Nature* (New York: George Braziller, 1988).

Ronan, Colin A. *Science* (New York: Facts on File, 1982).

Rose, Steven, and Lisa Appignanesi, eds. *Science and Beyond* (New York: Basil Blackwell Ltd., 1986).

Russell, Bertrand. *Wisdom of the West* (New York: Crescent Books, Inc., 1959).

Seldes, George, ed. *The Great Thoughts* (New York: Ballantine Books, 1985).

Serres, Michele, ed. *A History of Scientific Thought* (Cambridge, MA: Blackwell Publishers Ltd., 1995).

Simonton, Dean Keith. *Greatness* (New York: The Guilford Press, 1994).

Thomas, Lewis. *The Fragile Species* (New York: Charles Scribner's Sons, 1992).

Tipler, Frank J. *The Physics of Immortality* (New York: Doubleday, 1994).

Williams, Trevor I. *Science* (New York: Oxford University Press, 1990).

Zeldin, Theodore. *An Intimate History of Humanity* (New York: HarperCollins Publishers, 1994).

Index

Abbeville, France, 195
Academy for Experiment, 40
Academy of the Mysteries of Nature, 338
Abdera, Greece, 54
Academy of Lynxes, 31
Adams, Henry, 84
Admiralty Research Laboratory, 304
Alamogordo Air Base, 99-100
alchemy, 55-56
Almagest, The, 8
alpha particles, 70-75, 80, 90
alpha rays, 70-75, 80, 90
Alpher, Ralph, 168, 169
al-Sufi, 148
Altman, Sidney, 269
Alzheimer's disease, 325, 342
Amgen, Inc., 330-331
amino acids
    breakdown of proteins into, 262
    building blocks of life, 268-270
    creation of, 312-317
    creation of proteins from, 312-321
    genetic codes for, 315-321
    genetic diseases, 321, 324-325, 345
    number of, 312, 315, 320
Amorites, 32
Ampère, André-Marie, 165
Anaxagoras, 54, 55, 57, 68, 82
Anaximander of Miletus, 55
Anaximines of Miletus, 55
Andromeda Galaxy, 148-149
anode, 66
Archbishop of Pisa, 33
Aristotle
    elements, 53, 55, 57, 58
    Scale of Being, 186
    scientific method, 53, 338
    theory of heredity, 283

theory of universe, 5, 6, 8, 10, 17, 28, 38, 53, 146, 161, 162, 185, 339
theory of volcanoes and earthquakes, 243
writings, 54
Ascension Islands, 199, 252-253
Ascent of Man, The, 53
asteroids, 147, 158, 173, 177
astrology, 6-7, 23
astronomy, 143-170
    Brahe, Tycho, 13-16
    Galileo, 27-39
    Hubble, Edwin Powell, 149-154
    Kepler, Johannes, 17-23
    modern, 145-163
    radio astronomy, 164-170
Astrophysical Journal, 169
astrophysics, 153-159, 168-170
atherosclerosis, 326-327
At Home in the Universe, 280
atom, 51-100
    electron, 70-72, 74-82
    energy, 84-100
    leptons, 82
    mass, 86
    muon, 82
    neutrino, 82
    neutron, 80-82, 86
    nucleus, 65, 71, 75, 79-83, 87-100
    proton, 82, 86
    quarks, 82
    size, 79
    strong nuclear force, 87-90
    tau, 82
    valance of, 62, 79, 312
atomic bomb, 84-100
    chain reaction, 92-100
    Einstein, Albert, 94-96
    Roosevelt, Franklin D., 94-96

life, 14-16
nova, 146
scientific method, 15-16, 339
Branch Davidians, 172
breast cancer, 325
Brief History of Time, A, 345
British Association for the Advancement
of Science, 209
Brno, Czechoslovakia, 282, 283
Bronowski, Jacob, 53, 217, 337, 339
Brookhaven National Laboratory, 177,
347
Brown, Robert, 259
Bruno, Giordano, 24-26, 37
bubonic plague
Black Death, 3-5
Great Plague of London, 42
Buffon, George Louis Leclerc de, 189-
190, 194, 195
Burke, Bernard, 169

Caccini, Father Thomas, 32
Caesar, Julius, 9
California Institute of Technology, 307
Calvin, John, 12
Cambridge University, 42, 45, 80, 96,
197-198, 304
Campo di Fiori, 26
Canary Islands, 199
cancer, 324-325
Cannizzaro, Stanislao, 61-62, 79
Castelli, Benedetto, 32, 33, 34
catastrophism, 191-196
cathode, 66
cathode rays, 67, 68, 70
Cavendish Laboratory
Bohr, Niels, 76
Bragg, William Lawrence, 305-309
Cockcroft, John D., 91
Crick, Francis, 304-309
DNA research, 304-309
nuclear physics, 80
Oppenheimer, J. Robert, 96
Rutherford, Ernest, 73
Sanger, Frederick, 322
Thomson, Joseph John, 70
Walton, Ernest, 91
Cech, Thomas R., 269
cells, 118, 255-281

blood, 264, 283, 312, 316, 321
chloroplasts, 263
cytoskeleton, 262, 266
discovery of, 258
endoplasmic reticulum, 261
evolution of, 271, 275-277
Golgi apparatus, 262
lysosome, 262
macrophages, 278-280
membrane, 262-263, 267, 270
mitochondrion, 262
mitosis, 264-267
nucleic acids, 298-303
nucleus, 259-261, 263-267, 298-318
organelles, 260-263
reproduction of, 260, 264-267, 314
ribosome, 261
types of, 259, 276
Cell Substance, Nucleus and Cell Division,
264, 285
Cellular Pathology, 260
cerebral palsy, 329
Ceres, 158
Chadwick, James, 80, 91, 293
Chargraff, Erwin, 299, 318
Chargraff Ratios, 299, 309, 317
chemical bonding, 61-62, 79, 87, 269,
270, 301, 311-312
chemical elements, 61-66
chemistry, 55-66
Chernobyl nuclear reactor, 341
chloroplasts, 263
chromosomes
composition of, 289
discovery of, 264
division of, 266, 287
formation in cell, 266, 311-313, 316,
321-322
genes, 289-293
meiosis, 286-291
mitosis, 264-267
number of human, 264, 287, 312
mapping of genes, 290, 292, 326-327
pairing of, 287-291
recombination, 290-293
Church of England, 198
Ciba-Giegy Corporation, 331
classical physics, 44
cloning, 322, 328, 329, 338

About the Authors

David Eliot Brody has been writing professionally since 1976 and has been practicing law in Denver, Colorado, since 1974. He has written several books and articles on various subjects and has lectured extensively throughout the United States in his special areas of expertise in the law. He has also been a lecturer and part-time faculty member at the University of Colorado and other universities, teaching graduate and undergraduate courses in law, sociology, and political science.

He served on the board of directors of the National Writers Club, is a member and former president of the Colorado Authors League, and is a member of the National Association of Science Writers, MENSA, and various other organizations. He lives with his wife and two sons in Denver.

In the early 1980s Mr. Brody began an independent research project on the history of science and the role of science in modern society, and began developing the historical background and identifying the scientists involved in the greatest scientific discoveries. That project culminated in *The Science Class You Wish You Had.* As an accomplished researcher, writer, lawyer and professor—but not a scientist—he presents science conceptually and in a manner that nonscientists can readily understand.

Arnold R. Brody, Ph.D., David's brother, is professor of pathology and the director of the Lung Biology Program in the Department of Pathology & Laboratory Medicine at Tulane University Medical School in New Orleans. Formerly, he was an adjunct professor at Duke

University and head of the Pulmonary Pathology Group at the National Institute of Environmental Health Sciences (at Research Triangle Park in Raleigh-Durham, North Carolina), which is part of the National Institutes of Health.

His primary area of expertise is in cell biology, including molecular biology, chemistry, anatomy and related fields, and he is regarded as one of the leading experts in the world in the area of environmental lung disease and cell biology. He has written more than 125 scientific articles and 25 chapters in books on the findings resulting from his scientific research. He also lectures extensively all over the world. He is married and has two daughters.

A large number of Dr. Brody's articles and book chapters concern new discoveries in his field. Through this work, together with his lecturing, his participation on boards, committees, and at science conferences, and his strong interest in the history of science, he knows firsthand the extraordinarily complex process that typically leads to a significant scientific discovery. Thus, as reflected in *The Science Class You Wish You Had*, he has special expertise to write about the greatest scientific discoveries in history and the scientists who made them.